Corvettes

Corvettes
1953-1988

A collector's guide
by Richard Langworth

MOTOR RACING PUBLICATIONS LTD
Unit 6, The Pilton Estate, 46 Pitlake, Croydon CR0 3RY, England

ISBN 1899870 11 3
First published 1983
Reprinted 1985
Reprinted 1986
Reprinted 1987
Second edition 1988
Reprinted 1991
Third edition 1996

Copyright © 1983, 1988 and 1996 Richard Langworth and Motor Racing Publications Ltd

All rights reserved. No part of this publication may be reproduced, stored in a retrieval system, or transmitted, in any form or by any means, electronic, mechanical, photocopying, recording or otherwise, without the prior permission of Motor Racing Publications Ltd

British Library Cataloguing in Publication Data: A catalogue record for this book is available from the British Library

Printed in Great Britain by The Amadeus Press Ltd
Huddersfield, West Yorkshire

Contents

Introduction and acknowledgements			6
Chapter 1	Ancestors and parentage	America's first sports cars	9
Chapter 2	The first-generation Corvettes	1953 to 1955	20
Chapter 3	Restyling and refinement	1956 to 1962	27
Chapter 4	Early Corvettes in competition	1953 to 1962	45
Chapter 5	The Sting Ray era	1963 to 1967	53
Chapter 6	Later Corvettes in competition	1963 to date	71
Chapter 7	Stingray and a changing image	1968 to 1983	79
Chapter 8	Showcars and specials	The might-have-been Corvettes	101
Chapter 9	The New Generation	1984 to 1988	121
Chapter 10	Buying a Corvette	Model identification and collector preferences	141
Chapter 11	Corvette ownership	Clubs, spares and restoration	149
Appendix A	Corvette milestones 1953 to 1988		152
Appendix B	Technical specifications and major options		153
Appendix C	Serial numbers, production, weight and base prices		157
Appendix D	How fast? How economical? How heavy? Performance figures for Corvettes		158
Appendix E	Interchangeability of Chevrolet small-block engines		160

Introduction and acknowledgements

What, another book on Corvettes? You already need a small bookcase to house a complete Corvette library. But if you're reading this far you must have bought the book – so I'd better assure you that there's something in it for you.

Despite monumental histories, quick-and-witty lash-ups, picture books, monographs, restorer's manuals and a two-inch reprint, the Corvette has never had the unique treatment of the MRP *Collector's Guide* series. MRP have made a speciality of this style of publication, which is why they have so many imitators. The concept: a blow-by-blow account of the way a car developed; the design and engineering avenues down which it might have progressed but didn't; a look at its competition career, if it had one (this one definitely did); detailed appendices providing quick facts you'll need in a hurry; and that all-important chapter about selecting, buying and living with the vehicle in question. That's what we've done here, for all Corvettes from 1953 to the seminal 1984 model – which was surely the best bit of Corvette news since the Sting Ray arrived 34 years ago.

Indeed the '84 struck us as so good (and potentially collectable) that we couldn't leave it out. We have therefore added a chapter to this *Collector's Guide* that the others generally don't include. For those of you who joined so many others in disappointment when the 1980 model did *not* prove to be the Aerovette, pop open a bottle of your favourite bubbly, or wend your way to the nearest dispensary of Real Ale, and join me in a toast to its worthy '84 successor, and successive models through 1988.

The Corvette has had a chequered history, and a lot of downs and ups. Those privileged pickpockets who run our lives in Washington have done their level best to bundle America's only sports car up in a snaggle of regulations, admonishments and commandments designed to turn it, basically, into a two-seat Cadillac; and some Corvettes over the years have been pretty dim representatives of the breed. But the marque's history before the Feds began monkeying (and now, it seems, in recent years) is a lot different. Herewith I hope to provide you in one handy package all the statistics and identifying material you'll need, whether you own one, plan to own one, or just dream about the impossible goal.

Which reminds me to tell my American readers what the British think of Corvettes, and *vice versa*. The British – or at least the British I've met – think they're terrific – all of them! I am ever amazed when a native friend beside me, rocketing down the M1 motorway, manoeuvres himself within scanning distance of a rare GB Corvette, almost always the fourth-generation (post-1968) variety, and exclaims 'Crikey, what a motor!'. In vain I protest that the car is a wheezing 1978 sled, that it steers like a lorry, has no guts at all, rattles like a well-worn Morris Minor and is about as much fun to navigate as the *QE2*. All is lost in the admiring, almost quixotic British obsession with the swoopy, wide-track monster in the other lane. (The '84 model kept the swoop, handles and goes like hell.)

On the other hand, the typical UK enthusiast has less than a nodding acquaintance with the pre-1968 models – the quaint 1953/55 with its toothy grin and prong-horn tail-lamps; the sleek and fast 1956/62 with its potent V8 engine, either 'fuelie' or otherwise; and the magnificent and short-lived 1963/67, which has been the hottest thing in USA collector circles for years. I hope British readers will get to know these earlier cars better through these pages.

Maybe the grass is always greener. Just as Sir Winston Churchill is more universally admired in America than in Britain, the Corvette provokes vast division in the US. You had to live with Winston; we had to live with the Corvette. Sports car people this side of the pond either like it or loathe it. I have to admit that in my early days I was among the loathers. I drove British sports cars, and I knew as a surety that those glass-fibre widgets from Missouri were beneath contempt.

I remember as if it were yesterday the grassy hillsides of Thompson, Connecticut, home of a perilous road course with an uphill-downhill hairpin arranged to give pause even to Lotus Sevens. Thompson was a Corvette-killer. The big A-Production monsters were quick enough in the dry, but we'd pray for rain, so as to watch Bruce Jennings' Porsche Carrera truly trim their glass-fibre hides.

The last time I visited Thompson, now sadly defunct, Bob Grossman was manhandling a Sting Ray coupe through the twisties, the rain pelting down and the Carrera snapping at his heels, getting closer every lap. Each time Bob came round there was just a little less glass-fibre hanging on to his car – somewhere around the hairpin, he'd pranged something hard, and the fracture just kept expanding. By the finish, Bob was driving a somewhat streamlined chassis. We roared with delight! Gawd, did we hate those 'Vettes!

But as time went on and I began to write for a living, I also began to revise my judgments, and not because I was paid to edit or write Corvette books. Karl Ludvigsen's epic history, *America's Star-Spangled Sports Car*, was where I really began to appreciate the inner workings of the car and its makers, while serving as one of the book's editors 25 years ago. Karl taught me many things over the years, but the education he provided on the Corvette was one of the most compelling.

It's true that the car started off as the most unlikely sporting machine you could possibly imagine. And there were times when the Corvette solution to the problem of fast motoring was akin to slicing a tomato with a 30-pound axe. Brute force, that was the answer.

But by the 'Sixties, people like Zora Duntov, Bill Mitchell and Ed Cole had begun to get the message across that Chevy's sports car had to *handle*; that its lines should be functional, not faddish; and that above all it had to stop. They worked, as I hope this book reveals, a minor miracle. The Corvette had matured with the V8 engine in 1956 – but it didn't hit its peak until the Sting Ray of 1963. The '84 represented another pinnacle.

I wish to thank my late friend Sam Folz, of Kalamazoo, Michigan, who provided the entire chapter on seeking, buying and enjoying Corvettes. Sam was a man of broad and noble tastes – he ran one of the best T-series MGs in North America and restored one of the first Triumph TR4s imported here – but his first love was Corvette and his knowledge was encyclopaedic. A special note of thanks goes to Becky Bodner, of *Corvette News*, the voice of the marque, for helping me with most of the photos in the competition chapter. A third big 'thank you' to Chris Poole and Frank Peiler, of *Collectible Automobile*, for their assistance in my researching Chapter 11, and to Ed Lechtzin, of Chevrolet Public Relations, who answered my questions on the more recent models and helped me with production figures. Thanks also to Publications International for allowing me to quote an excerpt from my 1978 book on Corvette.

My own research on matters Corvette began with Ludvigsen, continued in Detroit on various related assignments over the years, and progressed through the beautiful Corvette publications of Michael Bruce Associates. Of critical importance to my archival source work were those recorders of the motoring scene: *Autocar, Road & Track, Car and Driver, Car Collector, Car Classics, Motor Trend, Sports Car Graphic, Automotive News, Corvette News* and *Vette Vues*.

At Chevrolet, at one time or another, or at the Tech Center, I have benefited from the assistance of the late Bill Mitchell, Chuck Jordan, the late Edward N. Cole and David Holls. I have quoted herein from interviews with Mitchell and Duntov conducted by my friend Rich Taylor, for a lengthy feature on Corvettes when I was editor of *Car Classics*. In addition to these, I must spread out a blanket thanks to all at Chevrolet PR, who have so many times assisted me on what must have seemed to them a fool's errand – and which had nothing to do with selling the current product. To all these, and any I have failed to mention, my very sincere appreciation.

It is important to acknowledge my personal UK connection, longtime friend, collaborator and colleague Graham Robson, of Burton Bradstock, Dorset, whose encouragement and assistance are like an extra hand on the end of a 3,000-mile-long arm; and to John Blunsden and the MRP staff, who have promised to translate the American into English, when not preoccupied with selling what's left of the Robson/Langworth *Triumph Cars* at a veritable bargain price (please buy that book next!).

Car enthusiasts owe a heavy debt to Chevrolet for its faith in the Corvette sports car. Corvette is really a very small item for a Division that makes millions of vehicles a year, and its end would cause little financial grief – in fact probably the reverse. But through it all they have persisted in their devotion to an idea, and today's team of Corvette engineers and stylists promises to continue delivering exciting sports cars in the years ahead.

RICHARD M. LANGWORTH

Hopkington, New Hampshire
June 1996

CHAPTER 1

Ancestors and parentage

America's first sports cars

According to the December 1982 issue of *Inc* magazine, one of the 500 fastest growing private corporations in the United States is Corvette America, Boalsburg, Pennsylvania. Dan LeKander founded the company in 1977 after 'totalling' his 1966 Sting Ray. Deciding to 'part-out' the car, he had sold $6,000 worth of it before he took stock and realized he still had an inventory of $4,000. Corvette America had begun. In 1981, it achieved $3 million in sales, and over its five years of life the business has grown 1,220 per cent.

The story is at once a fit comment on the status of Corvettes among today's collectors and a challenge to the writer of any new book on them. What, after all, can be said about the Corvette that hasn't been said before? The answer is, I suspect, very little from an historical standpoint. But the subject of collecting and living with Corvettes has yet to receive the concentrated blue light of the *Collector's Guide* series, and a look at the marque from this angle is, I suggest, very worthwhile.

Let us dispense first with hoary legend and unfounded but oft-repeated rumour — the Corvette was *not* America's first sports car. Nor was it America's first post-Second World War sports car. Nor, in fact, was it for very long America's *only* sports car; not if you count Nash-Healeys, two-seater Thunderbirds, Corvair Corsas, Camaro Z-28s, Firebird Trans Ams and the sensational new Chrysler G-24.

What Corvette *has* been is America's most successful sports car — and the most popular of that type among the half-million collectors that constitute the classic car avocation.

To chart the *affair d'amour* between Yanks and sports cars, however, we really must drop back to the palmy days before the 'War to End Wars', more accurately known in Britain as the Great War — that between 1914 and 1918. Those were the days when 'light two-seaters' from Simplex, Lozier, Locomobile, Stutz and Chadwick held sway — true sports cars all, capable of rapid motoring on both road and track.

Everything is relative, mind you, and 'roads' in those years were mostly muddy trails, while 'tracks' consisted either of sand in its natural state or swaying, elevated courses constructed of splintery timber planks. (Brooklands and Indy were, of course, notable exceptions.) Also, in pre-WW1 years the word 'light', in reference to sports cars, meant spartan and slightly less ponderous versions of the big open tourers, with a capacity for only two passengers instead of seven or eight. If you say it took a man to handle a Stutz Bearcat, you just might qualify more as an educated individual than a male chauvinist pig, because a Bearcat *did* require a man, and a brawny one at that.

The first really light or at least nimble sports car was the 1911 Mercer Raceabout. The breed survived through 1923, but the 1911-14 T-head versions are the best remembered. Driving one today reminds you somewhat of the early cart-sprung V8 Corvettes — light, responsive, not too comfortable and blindingly quick. Mercer was to Stutz as Bugatti was to Bentley (Ettore would, perforce, agree). Like the Corvette, the Mercer was tolerant of considerable modification; in 1912, for example, Ralph de Palma coaxed a Mercer Special up to 120mph (instead of the usual 80) with a 445 cubic inch racing engine.

The between-wars period in the USA saw a dearth of genuine sports cars. There were some good racers — the Millers and the early Duesenbergs — but these were not designed for road

America's first sports cars were the big, ponderous two-seaters exemplified by this 1908 Locomobile. Using heavy ladder chassis from conventional touring cars, a bonnet, two deep bucket seats, a petrol tank and a trunk were the basic 'body' features. 'Monocle' windscreens were the racing screens of the pre-WW1 era. *(Henry Ford Museum)*

A T-head Mercer Raceabout blasts through a turn at Point Breeze dirt track, Philadelphia, driven by prominent Mercer pilot Hughie Hughes. Despite its age, the Mercer is surprisingly easy to drive today. Clutchwork is light, and the non-synchro gearbox slips through the gears like a unit two generations younger. *(Henry Austin Clark Jr)*

motoring. There were 'sporty' cars, like the Cord 810/812 and Auburn Speedster, but they were more *art deco* GTs than dual-purpose machines. Real sports cars in the 1918-40 period were built and mainly enjoyed not by Americans but by the Britons, Italians, French and Germans.

After the Second World War a sports car renaissance began in America. GIs stationed in Britain, those of whom some of the locals chanted, 'over-sexed, over-fed, over-paid and over here', began returning home in 1945, and a goodly number brought an MG TC with them. The TC was the progenitor of a generation of sports cars that the average American could readily afford. (The old Mercers and Stutzes had really been priced far beyond the reach of the common man.)

The TC is remembered in Britain with a certain loyal nostalgia. In America it is regarded as nothing short of a legend. To the Yanks it was more than a car; it was a defiant expression of a whole new approach to motoring. Rank and file American cars of the late-'Forties, still a decade away from their period of elephantiasis, were by comparison a pretty dull lot. The late Ken Purdy called them 'a turgid river of jelly-bodied clunkers', and

Mercers were ready to race off the showroom floor, quintessential sports cars by any standard. Ready for the 'off' in California are the Raceabouts of Dick Bentle, the Los Angeles agent (car No 2) and the legendary Barney Oldfield (car No 3). Last Mercer Raceabouts were built in 1923, and the 1918-40 period was a spare one for American sports cars. *(Henry Austin Clark Jr)*

Frank Kurtis with his Ford-powered Kurtis-Kraft in California in 1949. Its flathead Ford V8, darling of hot-rodders in the immediate post-WW2 era, was easily modifiable. Kurtis dropped it into a strong chassis and body of his own design to produce a successful sports car. Earl 'Madman' Muntz bought out Kurtis in 1950 and produced similar cars called Muntz Jets through to 1955.

The Crosley Hot Shot (a 1951 model is shown) was a spartan rig powered by a 44 cid L-head four of 26.5 horsepower, though both engine and chassis responded favourably to modification. A basic Hot Shot cost only $950. This incredible midget won the Index of Performance at the Sebring 12 Hours in 1951, but Crosley Motors closed its doors in early 1952.

First Nash-Healey was the 1951 model, with divided windscreen and vertical Nash grille. This model used an aluminium body and was powered by an ohv Nash six tuned to deliver 125bhp by Donald Healey. Racing versions finished ninth in class in the Mille Miglia and fourth overall at Le Mans. Later (1952-55) Nash-Healeys used steel bodies designed by Pininfarina and inboard headlamps. Coupes as well as open models were offered on longer wheelbases for 1953-55. Total production was only 404.

wasn't far wrong.

TC drivers loved to bait the over-stuffed Buicks, Chryslers and Lincolns on American highways, tailgating them along the straights and then running around them on winding roads, generally giving their drivers extreme cases of heartburn. Flat-out at 75mph and the TC driver thought he was Nuvolari.

When William Lyons introduced his breathtaking Jaguar XK120 in 1948, another kind of sports car appeared in America at the upper end of the price scale, and interest in the breed burgeoned. By the mid-Fifties, cars like the Triumph TR2 and the Austin-Healey 100 had filled the middle-price gap, and the sports car stampede was in full force.

We should not, of course, over-rate the TC/XK phenomenon, because sports cars as a genre never comprised more than a radical fringe of the automotive scene in those days. In 1952, for example, there were only 11,000 sports cars registered in the entire United States of America. They represented 0.025 per cent

This dramatic looking Kaiser-Darrin DKF-161 was an early user of glass-fibre bodywork. Styling was by Howard Darrin, who had earlier handled most of the smashing 1951 Kaiser, and the taillights were from Kaiser's 1952 model. The dramatic sliding door was Darrin's most unusual feature. The body was dropped over a Henry J chassis and a Willys F-head 161 cid six, and the car's styling was far better than its performance. Only 435 were built, all designated 1954 models.

of total new car registrations — and 1952 was a low production year. Such statistics were not the stuff with which to excite the big Detroit manufacturers about building sports cars of their own, and the only domestic versions available through 1952 were produced by small and even tiny specialist manufacturers.

From 1949 through to its demise as a car builder in 1952, Crosley Motors, of Cincinnati, Ohio, had offered the spartan but entertaining Hot Shot and Super Sport. (The chief distinction was that the Super Sport had opening doors; you climbed into or out of the cheaper Hot Shot.) One such Crosley won the Index of Performance at the Sebring 12 Hours in 1951. Meanwhile, in California, Frank Kurtis built an envelope-bodied sports car using mainly Ford components, which was capable of being raced on a road course. In 1949 Kurtis was bought out by Earl 'Madman' Muntz, who continued the concept with the rather larger Muntz Jet.

The most successful American sports car from a competition standpoint was Briggs Cunningham's fleet range of two-seaters produced between 1951 and 1955. Cunningham saw a severe dichotomy between primarily-racing and primarily-touring sports cars. Among the former, the 1952 C-4R finished fourth at Le Mans and its successor, the C-5R, came in third the following year — beaten only by two D-type Jaguars. The touring types were represented by the $10,000 C-3, powered by a hemi-head

Late-night mods on the Excalibur J, another Henry J-based sports car, but one that could really go. Designer Brooks Stevens conceived the Lotus Seven-like Excalibur, which was a far cry from his baroque models of the same name in later years. The Willys F-head engine was expected to produce daylight-seeking pistons at 5,000rpm, but Stevens revved it to 6,500 without damage. *(Brooks Stevens)*

Proof of the pudding. An Excalibur J leads the pack (including several XK120s) at a Wilmot Hills SCCA Regional race in 1953. This photo was taken on the second lap! The second Excalibur J can be seen rounding the turn in the midst of the ruck. *(Brooks Stevens)*

Chrysler V8 of 180-plus horsepower and sold in roadster or Vignale coupe bodies. Fewer than 40 C-3s had been produced before Cunningham closed up. They had lost money for five consecutive years, and the tax man told Briggs he would necessarily be considered to be running a 'hobby' if he pressed on.

Moving up the scale towards the volume manufacturers, two smallish companies which produced cars of some influence on the future Corvette were Nash and Kaiser-Frazer. George Mason, the cigar-chomping President of Nash, met Donald Healey on a trans-Atlantic voyage, and by the time they'd arrived in New York the Nash-Healey concept was laid down. Open and closed cars were built between 1951 and 1955. They used a tuned version of the ohv Ambassador six, and were most impressive in competition; a prototype finished fourth at Le Mans in 1950, and a stock Nash-Healey ran sixth overall and fourth in class at the 24 Hours in 1951. In 1952, when Cunningham was placed fourth, a Nash-Healey finished third overall, behind two far more powerful Mercedes-Benz racing cars. Unfortunately, very low sales and the Nash-Hudson merger, followed by the death of Mason and the decision to concentrate on economy Nash/Hudson Ramblers, ended the Healey project by 1955.

Kaiser-Frazer's compact Henry J chassis spawned two very different sports cars, the Darrin and the Excalibur J. The former was a glass-fibre two-seater with novel sliding doors, conceived by Howard Darrin, the author of the outstanding 1951 Kaiser saloons. The Kaiser-Darrin DKF-161 went into production as a 1954 model, but a declining company and feeble sales effort saw only 435 models produced. Whether or not the Darrin was a true race-and-ride sports car is questionable, since it was not very successful in competition. A few did race, however, and Mrs Briggs Cunningham did very well in her performances on Sports Car Club of America road courses.

Brooks Stevens' Excalibur J was pure sports car. Like the Darrin, it was constructed on the primitive Henry J chassis, and used an overhead-valve Willys-Overland engine. But Stevens planned to sell the Excalibur J for about $2,000, whereas Darrins listed for $3,700 — and Stevens had carefully designed his car to race and win. In SCCA contests, Excalibur Js were often found running away from Jaguars and Ferraris — which was almost as astonishing to Stevens as it was to the spectators. But the Excalibur J project fell apart, as did the Kaiser Darrin when K-F ran into financial problems around 1953.

The Nash-Healey and Kaiser-Darrin, if not the other nascent American sports cars mentioned above, were well known and appreciated by certain second-level executives at the Big Three

The earliest user of a Glasspar body was the Woodill Wildfire, a Willys-based special produced in both finished and kit form from 1952 to circa 1956. Wildfires used a standard Glasspar body which was also made available for Ford chassis. Several hundred were marketed, mostly in kit form. The gullwing dash was inspired by the contemporary MG. The builder, Woody Woodill, even offered a children's model, dubbed the Brushfire.

manufacturers, notably Virgil Exner at Chrysler, Frank Hershey at Ford, and Harley Earl and Ed Cole at General Motors. The Kaiser-Darrin, which had been previewed as early as 1952, was arguably the most interesting, as it shunned a steel body for glass-reinforced plastic (GRP). Glass-fibre had certain advantages for a low-production sports car; tooling costs were minimal by comparison to steel stamping dies, and the low amortization costs could be absorbed over a small production run. Sports cars *per se* didn't excite the top-level managers of the Big Three, but GRP did. All three companies at one time considered across-the-boards shifts from steel to GRP bodies for their regular lines of cars.

The leading fabricator of glass-fibre bodies was Bill Tritt, who founded the Glasspar Company in California in 1950. Glasspar built a good selling line of small boats, along with exotic looking two-seat bodies for American car chassis like Crosley's and Ford's. They were the 'replicars' of the Fifties, although they didn't replicate anything — they improved on the conventional. Tritt then convinced the US Rubber Division in Naugatuck, Connecticut, of the practicality of GRP in automobiles. In 1952 the two companies joined to produce bodies for 100in-wheelbase cars at $650 each.

One day in 1952, US Rubber management received an invitation to visit Detroit. Chevrolet was interested in a GRP automobile body for a two-seater, and Chevrolet was going to need it quickly.

Surprisingly enough, the initial interest in GRP originated not among the engineers but among the designers — led by Chief of GM's Art & Colour Studio, Harley Earl. Long ensconced at GM, Earl had joined the company after freelancing the stunning 1927 LaSalle. Immensely talented himself, he attracted dozens of bright people, including Bill Mitchell, Virgil Exner and Frank Hershey, all of whom began their design careers at Art & Colour. By 1950, Harley Earl's GM styling team were the leading purveyors of automotive design in Detroit; where they led, the other companies would follow.

Harley Earl had closely observed the success of Glasspar and was acquainted with Bill Tritt. He also had met and talked to Howard Darrin, whose first experimental glass-fibre automobile was turned out as early as 1946. And — mark this — Harley Earl was a car fan. He liked fast cars, particularly sports cars; he attended races, both Stateside and in Europe; and he greatly admired Bill Lyons' Jaguars and wished GM could build something comparable. Among all those who influenced the development of the Corvette, Harley Earl was the most important.

Very close behind comes Edward N. Cole, Chief Engineer

An artist's renderings of the Motorama Corvette. Obvious differences from production cars were the minimal side moulding and the downward pointed fin. Pushbutton door releases were also eliminated in production. The Motorama car featured a hydraulic system which automatically opened and closed the bonnet and doors as the car revolved on a turntable. This Corvette toured with the GM Motorama in the early months of 1953 and received a tremendous reception.

(later General Manager) of Chevrolet, destined to become GM's most charismatic and imaginative President. Like Earl, Cole wanted a GM sports car; but more specifically he wanted it to be a Chevrolet. Since arriving at the Division in 1951 Cole had been horrified by the old-hat image of Chevrolet and was determined to take off after Ford in the performance field. A sharp sports car was the quickest path down this road.

Working in close liaison with Cole and Chevrolet, Harley Earl's stylists began with two experimental projects in 1952: a conventional Chevrolet convertible with a GRP body, and a two-seater sports car. The latter was to be designed from the ground up, but to use as many off-the-shelf components as possible.

The convertible was built to test the durability of GRP — and to placate higher brass who had visions of a whole generation of full-size glass-fibre GM automobiles. Its strength was proven at an early stage. Driving on the GM Proving Grounds, a driver accidentally rolled the 'plastic' convertible at high speed. He escaped unhurt, and the body structure remained sound — the doors, boot and bonnet opened and closed without difficulty, despite a terrific pounding.

The sports car project was turned over to a young engineer fresh from the California Institute of Technology, Robert McLean. Working from the rear forward (counter to the norm), McLean placed the two seats just ahead of the rear axle, then positioned the engine and gearbox as far back as he could. The objective was a 50-50 weight distribution. McLean also lowered the drivetrain to lower the centre of gravity, and established the wheelbase at 102in — identical to that of the Jaguar XK120 (and not quite by accident — GM was testing and tearing apart 120s in the process of designing the Corvette).

17

A production model 1953 Corvette. The bonnet was hinged at the front for improved engine access, as Chevrolet had purposely set the powerplant far back in the chassis for optimum weight distribution. The soft top was manually operated and side curtains were used in lieu of roll-up windows. Note the shadow-box treatment of the number-plate and the dual exhausts exiting inboard from the rear body.

Budgetary constraints required the use of stock drivetrain components, and Ed Cole's Chevrolet V8 was still years away. There was only the 235 cubic inch Chevrolet Six, which produced an anaemic 105bhp in standard tune at the time. Working against deadlines with almost backyard methods, the engineers gave this eminently ordinary engine the best speed shop accessories they could — a high-lift, long-duration camshaft, solid instead of hydraulic valve lifters, dual valve springs and a modified cylinder head with 8:1 compression instead of the stock Chevy's 7.5:1. Water pump flow was increased and the pump itself was lowered, so that the large, four-blade fan could clear the anticipated line of the bonnet.

Long, hard labour was applied to the induction system, which was the most seriously modified component of the new Corvette's drivetrain. A special aluminium intake manifold was adopted to take three Carter type YH sidedraught carburettors. Contrary to rumour, these carbs were not designed specifically for the Corvette; the Nash-Healey had already used dual YHs. However, though Nash had been able to make its set-up work on automatic chokes, Chevrolet tried, failed to obtain choke synchronization and switched to manual choke for its production models. The result of these and other engine changes was an impressive increase in power to 150bhp at 4,500rpm — and the safe redline was a good 1,000rpm above that.

Chassis components were similarly pirated from spares shelves and modified where possible. Even so, the Corvette chassis was unique unto itself. For the first time in Chevrolet history, for example, Hotchkiss drive was used instead of traditional torque tube drive. Rear springs were conventional leaf type, but mounted outboard of the main frame rails for improved stability. Up front, suspension design was similar to, but not identical to, that of Chevrolet passenger cars. There were special spring rates and shock settings and a unique stabilizer bar; steering was GM-Saginaw recirculating ball, but its ratio was quickened to 16:1.

Great controversy has always surrounded the original Corvette transmission, Chevy's automatic Powerglide. This unit was foisted on the car by time limits, but in fact it offered certain advantages. Essentially unchanged from its passenger car form, it was rugged and well proven over millions of miles in the field. Unfortunately, it was hardly the kind of gearbox that enthusiasts appreciated, however, and it would remain a bone of contention.

The decision to use GRP rather than steel for the Corvette body came fairly late, despite Harley Earl's confidence in the stuff. Ultimately, the need for rapid production controlled the decision as much as the merits of GRP itself. Experimentation with the GRP-bodied Chevy convertible underlined its basic

The production model 1953 Corvette adopted a full-length bright metal side moulding. Cars were finished only in Polo White with red interior during the initial model year; whitewall tyres and a 150bhp Chevy six with Powerglide automatic transmission were standard. The 1953-55s had no exterior door handles.

strength in automotive applications. One factor that wasn't considered in those days before government regulations was the fire hazard; the early GRP bodies would burn to the ground in a matter of minutes.

The name 'Corvette' came from the fast naval vessel of that title — in a way it was a nod to the fact that there were far more GRP bodies in the water than on the road! It was a good name, and it stuck. A fastback special displayed in 1954 adopted the name 'Corvair' by combining 'Corvette' and 'Bel Air' (Chevrolet's top-line passenger model) — and this later appeared as the production rear-engined Chevy in 1960.

The target date for the debut of the prototype Corvette was the GM Motorama — the annual travelling extravaganza of cars, girls and gladiolas, scheduled for early-1953. Though the prototype was one of the few showcars that evolved in production with hardly any significant changes, some details were, of course, revised after the debut. Under the bonnet, the Motorama Corvette boasted a lot of chrome plated bits; it also had a shrouded fan and pancake air cleaners, which disappeared in production. The showcar's cowl-mounted fresh air scoops were dropped, too, though they reappeared on the 1956 Corvettes. The showcar had a hydraulic servo which opened and closed the bonnet and doors as the car revolved on its Motorama turntable — this, too, was dropped in production.

Both the Motorama Corvette and early production models used a one-person, manually-operated soft top, which disappeared below a trap door mounted behind the twin bucket seats. The car was equipped with bright metal-bound, removable Plexiglas side curtains and push-pull ventwings. The Motorama Corvette had exterior pushbuttons for door opening, but these were dropped on production cars. To open their doors you thrust your hand through a ventwing and moved an inside catch, rather like the early Triumph TR2 and TR3s and the Austin-Healeys. This thoroughly un-GM-like arrangement was eliminated forever on the 1956 models.

Gleaming in Polo White with a Sportsman Red vinyl interior, the Motorama Corvette was enthusiastically received by the public and by motoring writers at the 1953 Motorama. Chevy dealers were soon besieged with customers waving cheque books, asking the questions: 'When can I buy one of these cars?' and 'For how much?'. The answers proved to be: 'By mid-summer 1953' (if you were important enough) and 'About $3,500' (more than an MG, but far less than a Jag). Though the Motorama Corvette was still very much a question mark in the minds of his bosses, Harley Earl himself had no doubt. 'I knew damn well we'd build it', he said later, 'and by God we did!'.

CHAPTER 2

The first-generation Corvettes

1953 to 1955

The first production Corvettes were produced at an assembly line only six cars long at Flint, Michigan. Car number 1 was built on June 30, 1953, after a frantic, six-month long effort by the Chevrolet Production, Purchasing and Engineering Departments. Engines were shipped in from the Chevrolet engine plant in Tonawanda, New York, and frames had been acquired from an outside supplier.

The delay was largely caused by the GRP body. Forty-six separate body components were required to build each car. Producing them, at the Molded Fiber Glass Body Company in Ashtabula, Ohio, was at first a complicated affair, resulting in crude products and mixed quality. Formed on wooden jigs, the panels were 'glued' into full bodies. Production tooling was not available and GRP production techniques in those days were primitive. There is evidence on early cars of glass cloth hand-layout technique, whereas the later production models used more glass mat construction, and after the early units wooden dies gave way to steel.

The National Corvette Restorers Club has noted widespread evidence of non-conformity on 1953 models. For example, early cars were not equipped with Guide Y-50 outside rear-view mirrors, which became standard on later '53s and would remain part of the package through to 1962. The first 25-odd cars used snap-on full wheel covers from the stock Chevrolet Bel Air passenger cars — later Chevy would switch to Motorama-type wheel covers with dummy knock-off hubs. All 1953 models were identical in using Delco signal-seeking radios and recirculating hot-water heaters. All also featured complete needle instrumentation, including a 5,000rpm tachometer which notched up total engine revolutions on an odometer-like barrel gauge. This unique feature was retained on all Corvettes through to the 1959 model.

Because it was the first of the marque — and because it numbered only 300 units — the 1953 Corvette ranks among the most desirable models, and 1953 values usually lead those of all others, even the rare fuel-injection cars of the later 'Fifties. In recent years, some evidence has been produced of 'forgeries' — 1954 Corvettes modified to become '1953s'. The services of experts, particularly from the National Corvette Restorers Club, should be sought if the collector is in any doubt, but there are certain telltale signs which the 'forgers' sometimes overlook.

Chevrolet always placed a number, corresponding to the body serial plate, on the Corvette frame, but on 1953 models this number appears twice. It can be found with the aid of a torch and a small mirror without removing the body from the frame. Another feature common only to 1953s was the petrol line, which exited from the bottom of the fuel tank outside the right-hand main frame rail. Certain early models were equipped with wheel cover hubs mounted at 90 degrees to the standard location. All 1953s had two inside bonnet releases, mounted at the extreme lower corners of the dashboard — though this feature was carried into the very early 1954 models as well. The first 175 cars in model year 1953 used foot-activated windscreen washers; after car number 175 this was replaced with a vacuum system controlled by a dashboard press-button.

Enthusiasts seeking an early Corvette should not, on the other hand, be over-concerned if the car lacks the stock stainless steel headlamp stoneguards or the clear Plexiglas cover over the inset rear number-plate. These items were removed by law in certain states where they were not in conformity with motor vehicle laws. (Unlike the UK, motor vehicle regulations are not uniform in the USA, and each state has its own set of standards and inspection codes. Efforts have been made to render these laws reasonably uniform, but it wasn't until 1956, for example, that all states agreed on standard 6 × 12in number-plates!)

Ed Cole and Harley Earl didn't quite get their wish with the 1953 Corvette; it wasn't, after all, a Jaguar-beater. Nevertheless, it was a far better car than most reviewers said it was — then and now. Despite its slushy automatic transmission and mildly modified six-cylinder engine, it would sprint from zero to 60mph in about 11 seconds and exceed 'the ton' in top speed. If the '53 was not exactly competitive in its displacement class, it was as good an over-the-road light tourer as many of its longer established rivals. 'Chevrolet has produced a bucket-seat roadster that will hold its own with Europe's best, short of actual competition', noted *Motor Trend,* 'and (is better than) a few imports that cost three times as much.'

GM's upper management had decreed a run of only 300 1953 Corvettes because they were still not convinced that the new car was a viable volume offering. Despite tremendous enthusiasm and strident demands from dealers, they stuck to that limit, resulting in what today is a very rare car. Well over 200 of them have been accounted for in the years since; cars numbers 1 and 2 are believed to have been destroyed, but other early serial numbers are known. Yet the survival percentage of the 1953 Corvette is fairly low by comparison to its peers. The nearly-as-rare Kaiser Darrin, for example, has an 80 per cent survival figure. One reason for this could be that most 1953 models were carefully channelled, not to demanding enthusiasts, but to prominent people both within and without Chevrolet — in order to obtain their praise or opinions. Such owners are less likely to maintain and hold on to a car than the devoted fans who were demanding Corvettes at that time.

After 15 of the 1954 models had been assembled at Flint, Corvette production shifted to a specially constructed assembly line in St Louis, Missouri, and by mid-1954 this factory was producing the cars at the rate of 50 per day. Few major changes occurred on the first 1954s, but running alterations were made as time passed. Immediate identification of a '54 (assuming no

'What, me worry?', or 'Why is this man smiling?'. One of the first production 1953s, built at Flint, Michigan. There was many a slip before production got underway, and even when it did only 300 1953 models were produced. GM top management took a look and began to wonder about the wisdom of the whole thing.

changes have been made) is provided by the soft top — finished in tan instead of black as on 1953 models. The '54s also relocated their petrol and brake lines inboard of the main right-hand frame rail. Engines had a new rocker cover, the wiring harness was neater and it used plastic-insulated rather than fabric-insulated wires. Engines, built at Flint, carried the number suffix F54YG. Finally, a wider variety of colours were offered.

These are only the chief glass-fibre body components. Altogether there were 46, and it took a while for Flint (and later St Louis) assemblers to get them all to 'cure' and fit properly.

Though about four out of every five 1954 Corvettes were finished in Polo White, as per 1953, some 16 per cent of the cars came in Pennant Blue, with tan interiors and a dark tan lower dashboard and steering column. These and the white cars used red-painted wheels. Three per cent of production came in Sportsman Red, with a red-and-white interior. A very few were painted black, with the same interior as the red cars. Total production for model year 1954 was 3,640 units.

The first 1954s appeared with single-handle bonnet releases. At car number 1,900, the triple bullet-type air cleaner was replaced by a twin-chambered type. About one 1954 car in five was equipped with chrome-plated rocker covers and/or chrome-finished ignition shields — these modifications fell between serial numbers 1,363 and 4,381. At car number 3,600 the top mechanism was modified, receiving a new shape allowing the top irons to be mounted between the body and seat back — previously they poked through slots in the mouldings behind the seats.

Sales of the 1954 Corvette were lacklustre by GM standards, and midway in the model year a discussion ensued as to whether or not the car had a future. Those in favour of dropping it argued that the Corvette had proven neither fish nor fowl; it was more boulevard tourer than out-and-out sports car, yet it failed to provide boulevard-style comforts like roll-up windows, fresh-air heater (its heater was a primitive recirculating affair like a TR2's), and an automatic top or optional hardtop. The sports car people within GM disapproved of the automatic transmission and non-functional gimcrackery, like the dummy knock-off wheel covers. Dealers complained about chronic water and dust leaks, which

Initial cars were powered by a modified Chevrolet in-line six, tuned to deliver 150bhp (the standard passenger car version had 105). By 1955 the switch to V8s was almost complete, and only about six to ten of the '55s had sixes.

inch V8, producing 195bhp at 5,000rpm while weighing about 30lb less than the six. The V8 reduced the Corvette's 0-60mph time to below 9 seconds and brought its top speed to almost 120mph. Unfortunately, despite a banner year for the American auto industry at large, the Corvette simply refused to sell. Model year production was a mere 700 units.

It has been remarked that the 1955 showing made Chevrolet question again the continuation of the Corvette, but new-model lead times have confused the issue. The doubts had been considered and rejected in 1954, not 1955. Even as the few 1955 models trickled out of the St Louis works, Chevrolet had already made the decision to come out with a completely revised car for 1956, from which — as we all know — the Corvette never looked back again. The 1955 model was a mere stopgap.

seemed impossible to fix. It is hard to believe today, but as the 1954 model year ended in September, the St Louis production line had ground to a stop, and over 1,500 Corvettes were unsold on dealer lots.

GM Styling had proposed an egg-crate-like grille for the 1955 model to mimic that of the production 1955 Chevrolet. They also wanted a functional bonnet air scoop, a modified boot lid and relocation of the exhaust outlets (which were found to be soiling the surrounding finish). None of these ideas were allowed, however, because the debate about whether to build a '55 Corvette at all was raging. Ultimately, GM decided to give the sports car 'one last try'. As so many times in their past, GM managers made exactly the right decision.

Though a new, hotter cam gave the six-cylinder 1955 Corvette 5bhp more, fewer than 10 cars were built with this engine. The rest of the '55s carried Chevrolet's new and outstanding 265 cubic

The three-speed manual gearbox arrived late in the 1955 model year — very few '55 Corvettes had it. From 1956 onwards, though, 80 to 85 per cent of the cars were equipped with manual transmission. Also visible here is the 1953-57 centrally mounted rev-counter and flanking auxiliary gauges.

Even weight distribution was achieved by setting the engine well back on the chassis, which was designed by Robert McLean. The drive-train was installed as low as possible for optimum centre of gravity.

Body meets chassis at the St Louis works. This factory was specially constructed for Corvettes and assembled all cars after the initial 15 from Flint.

Generally speaking, the 1955s were smoother-looking cars, slightly thinner in section than their predecessors, and much more carefully put together. Most of the early production bugs had been removed and the product — although in the last year of its styling cycle — was better than ever before. The Pennant Blue colour option was soon replaced by Harvest Gold (with medium-green interior and dark-green top); Metallic Corvette Copper was a new colour for 1955, and Gypsy Red replaced Sportsman Red (with white vinyl interior and red saddle stitching and tan carpet and top). V8 models, which is to say literally all 1955s, had a small gold 'V' overlaid on the letter 'V' in the side script to signify the new V8 engine.

While six-cylinder 1955s retained a 6-volt electrical system, the V8s received the 12-volt system and their rev-counters now were calibrated to 6,000rpm. V8s also had electric windscreen wipers, rather than the vacuum type, and retained foot-operated washers. The only transmission listed was still Powerglide, but in 1955 the vacuum modulator was eliminated, allowing kickdown gear to be governed solely by speed and throttle position. However, very late in the model year, an unknown but small number of cars did receive a new close-ratio three-speed gearbox with a floor-mounted shift lever. Stick-shift models used a somewhat taller rear axle ratio — 3.55:1 instead of 3.70:1.

Among these first-generation Corvettes, values of cars in comparable condition are what one expects from the relative production figures. The 1953s lead all the others by a long way;

This very early 1953 Corvette sports early production wheel covers, smooth with a Chevrolet 'bow tie' emblem. Later, a special cover with dummy knock-off hub was adopted.

A cavalcade of Corvettes on Chicago's Lake Shore Drive in 1953. Ken Purdy's 'turgid river of jelly-bodied clunkers' is passing by on the other side of the road. By 1954 there would be several such seas of Corvettes in distributors' hands as the early cars sold in very low quantities.

the 1955s are positioned roughly midway between the '53s and '54s. For this writer's money, however, the saving one makes in opting for a 1955 Corvette over a 1953 is worth considering. The 1955 was a better car in every way — more fastidiously assembled, available in a wider variety of colours, and equipped with the all-important V8 engine. Nor does selecting a V8 mean one has to give up economy, for the 265 was a remarkably frugal engine with its bags of torque and low-rpm cruising speeds. Of course, we are dealing with a very esoteric range of models. None of them is inexpensive today. None of them ever will be.

The 1954 Corvette passes a somewhat earlier conveyance. The manually operated soft top dropped into a well under the hinged cover behind the seats. Wraparound windscreens were *de rigueur* in the USA in the mid-Fifties.

The 1955 Corvette with a V8 engine and a modified 'Chevrolet' script to so designate it. The 1955 model was the best-put-together of the first-generation cars, somewhat thin in section, and available in a wide range of colours.

CHAPTER 3

Restyling and refinement

1956 to 1962

There was no doubt in Chevrolet General Manager Ed Cole's mind that the Corvette had needed a V8 engine — as did the full line of Chevrolet passenger cars. Work on what became the 1955 V8 began as soon as Cole was installed at Chevrolet Division. With it, the Corvette was transformed from an acceptable if unspectacular sports car to a genuine performance machine. It remained now for Harley Earl's designers to upgrade the package. This they did in the second year of the Corvette V8, 1956.

The styling basis of the 1956 Corvette can be found in two 1955 Motorama showcars, the Biscayne and the LaSalle II (see also Chapter 8). The compact four-door pillarless Biscayne sedan contributed the vertical-bar radiator grille and concave body side moulding; the LaSalle II (which was shown in both roadster and saloon versions) moved the Biscayne's side moulding from the rear of the car to the front. Harley Earl later admitted that the side styling was really borrowed from the Classic-era 'LeBaron Sweep' — a notable hallmark of the LeBaron Coachworks and prominently found on Duesenbergs. Whatever its origins, the idea broke up the rather slab-sided bodywork and added dash to the package.

The actual 1956 production Corvette was a much more refined car than either of its showcar forbears. It also harkened to certain contemporary European sports cars, in particular the Mercedes-Benz 300SL. Historian Karl Ludvigsen has written that the 300SL 'was responsible for the 1956 Corvette's forward-thrusting fenderlines . . . and the twin bulges in its bonnet'. But a hotchpotch of different design elements the '56 was not; its overall shape was smooth and pleasant, with curves in the right places and very little ostentation other than the glittery grillework. Few traces of the less-functional 1953-55 generation remained — the dummy air scoops set atop the front wings and the fake knock-off wheel covers were the only elements enthusiasts could now criticize. In retrospect we can also say that the 1956 model (and its lookalike 1957 successor) were cleaner looking cars than the 1958s which followed. This good styling — and certain mechanical improvements — has made the 1956-57 Corvettes among the most popular and costly models for collectors today.

Along with improved function came considerably enhanced ergonomics. In 1955, Ford had sold five Thunderbirds for every Corvette, and Chevrolet product planners concluded that many of the lost sales were due to dated and clumsy design features on the 1953-55 models, such as the side curtains, the lack of exterior door handles, the primitive (for Detroit) soft-top and the poor visibility. So the '56 was fitted as standard with roll-up windows and door handles, and the option list contained a handsome, fully-upholstered lift-off hardtop with broadly wrapped glass and thin pillars. Unlike the Thunderbird (Ford had already decided to drop the two-seater in favour of a four-seater by 1958), Chevy shunned styling gimmicks like 'porthole' hardtop windows, exterior spare tyres and that apparition of late-'Fifties Detroit — tailfins. As the Thunderbird marched down the road towards pillow-soft ride and fashionable fripperies, the Corvette remained true to the dual-purpose character of the genuine sports car.

Though the '56 retained the 265 cubic inch V8 of 1955, a new high-lift cam developed by Corvette engineer Zora Arkus-Duntov and 9.25:1 compression combined to produce more horsepower — 225 at 5,200rpm, with 270lb/ft of torque at 3,600rpm. The

With an all-new body for 1956, the Corvette was transformed into a purposeful looking car bereft of the styling cliches of the 1953-55 series. Vertical grille teeth remained a trademark.

standard three-speed manual gearbox accordingly was hooked to a stronger clutch, which used 12 heat-treated coils instead of the previous diaphragm spring. Standard axle ratio was 3.55:1, though a 3.27 gave the car more legs; Powerglide could be supplied with the 3.55 axle for $189 extra.

The result of this tinkering was a typical 0-60mph time of 7.5 seconds and a standing-start quarter-mile in 16 seconds and at least 90mph. The big mechanical problem in 1956 was the brakes — still cast-iron drums with only 158sq in of lining area. Fade was apparent after one hard application. The '56 was, however, a good handler, with nearly perfect (52/48 per cent) weight distribution and quick, positive steering with only 3.5 turns lock-to-lock.

The Corvette for 1957 looked much like its predecessor, but evolutionary changes were made under the skin. The most important of these was a larger, 283 cubic inch V8 engine. The most interesting was Ramjet fuel injection, which permitted Chevrolet to claim one bhp per cubic inch. The injection had been developed at Rochester Carburetor Inc, though to a General Motors design. It was an exciting development, and 'fuelie 'Vettes' have become among the most desirable and pricey examples of the breed. But in practice, the fuel injection was not without problems — and these are worth considering if you plan to drive your future Corvette any kind of serious mileage.

The Ramjet design incorporated a special manifold, a fuel meter and an air meter; the latter directed the air to the various intake ports, where the fuel meter injected the precise amount of petrol required. A high-pressure pump, driven off the distributor, supplied fuel. The upper casting of the aluminium manifold contained the air passages and metering system, while the lower casting contained ram tubes and a cover for the top of the engine. Though a little exaggeration cannot be discounted, Chevrolet

Concave side sculpture came from the LaSalle II and Biscayne showcars. Cars could be had with this section painted a second colour, but they looked better in monotone. The exhaust exited through the rear bumpers, but not without soiling the nearby bodywork.

claimed that the result in brake horsepower was 283 at 6,200rpm. If so, it was the first time that a mass-production automobile engine had produced one unit of bhp for every cubic inch of displacement (cid).

Ramjet looked like the answer to racing drivers' prayers, and 'fuelies' soon took to the circuits — but the system proved unreliable. Racing mechanics had to block out the fuel cut-off to avoid a flat spot during hard acceleration. Worse, the fuel nozzles were prone to failure through dirt deposits and they absorbed heat, causing rough idling. Similar difficulties attended injected Corvettes destined for road use, and the result was that only 240 of the 1957 models were ordered with Ramjet. On the plus side, the fuel-injected Corvettes had awesome acceleration when they were running right — 0-60mph times averaged 6.5 seconds.

More important historically was the new 283 V8, produced by boring the 265 one-eighth of an inch to 3.875 inches. More versatile than the 265, the 283 could be had in five stages of tune — 220bhp with a single-barrel carburettor, 245 or 270bhp with dual four-barrel carbs, and 250 or 283bhp with fuel injection. Compared to the 265, the 283 V8 had longer-reach spark plugs, larger ports, wider bearings and oil-control piston rings. Dual exhausts, when equipped, were joined by a cross-over pipe, which equalized the exhaust flow through each muffler for even distribution.

In May 1957, the Corvette took another step towards becoming the complete sports car; a Borg-Warner four-speed gearbox (designed by Chevrolet), was added to the option list at $188. Actually this was a modified three-speed, in which reverse gear shifted into the tail shaft housing to make room for the fourth forward gear. The four-speed boasted close ratios — 2.20, 1.66, 1.31 and 1.00:1. The usual plethora of final drive ratios was offered. With a stump-pulling 4.11 axle the fuel-injected Corvettes were recording 0-60mph in 5.5 seconds and standing quarter-mile speeds of close to 100mph — even with this ratio, the fuel-injected 283 had a top speed of 130mph-plus. If this wasn't enough there was also a 4.56:1 ratio!

Important to the enthusiast driver, then and now, was RPO (Regular Production Option) 684, the Corvette 'handling' suspension, first offered on the 1957 models. This included a front anti-sway bar and heavier springs, heavy-duty rear springs

A lift-off hardtop was a Corvette option commencing with the 1956 models. Thin roof pillars were adopted to ensure good all-round visibility.

This is a retouched photograph, which had been taken of a car fitted with a duotone white hardtop and narrow-band whitewall tyres, neither of which appeared in production.

It is best to have a second person on hand whenever you want to either fit or remove a Corvette hardtop. This is a 1956 model.

and shocks, ceramic-metallic brake linings with ventilated finned drums, Positraction (limited-slip differential) and a quick-steering adapter that reduced the lock-to-lock turns from 3.5 to 2.9. With factory equipment like this, Corvettes were suddenly a competition force to be reckoned with. At Sebring in 1957, a pair of suitably optioned cars finished 1-2 in the GT class, 20 laps ahead of the nearest Mercedes, and 12th/15th overall despite some very potent Prototype-class entries.

The Sebring performance was not only a shot in the arm to the enthusiasts at Chevrolet Division who wanted a genuine sports car — it marked an awakening for the worldwide automotive press as well. 'Before Sebring, where we actually saw it for ourselves, the Corvette was regarded as a plastic toy', said one European motoring writer. 'After Sebring, even the most biased were forced to admit that the Americans had one of the world's finest sports cars — as capable on the track as it was on the road.'

Nineteen fifty-eight is remembered in the American motor industry as something slightly worse than 1982 — a really terrible recession year. Among the volume nameplates, only Rambler saw increased production over 1957. Among the speciality cars or 'sub-models', however, business was good. The all-new four-seat Thunderbird had record sales, and the not-so-new-but-still-pretty-good Corvette recorded a volume increase of close to 50 per cent. The model year figure was 9,168 units (compared to 6,339 in

The 1957 Corvette understeered on hard cornering, but roadholding was good despite considerable body roll. Few changes were made from the 1956 package, but sales improved.

1957 and 3,467 in 1956). But the '58 was in general a less exciting affair than the hotter versions of the '57. GM had joined with the other Detroit manufacturers in endorsing the recommendation of the Auto Manufacturers Association that companies should not take part in or sponsor racing cars and/or drivers. It was an ostrich-like position sponsored by dim-wits at the National Safety Council, who somehow imagined that racing encouraged highway mayhem. Some years later the same kind of reasoning created the infamous seat-belt interlock, which prevented you from starting your car until you buckled up — even to reverse down your driveway. A funny lot, we Yanks!

Anyway, for a time, GM and the rest toed the mark and kept out of the competition business. They also allowed this decision to work its way into the design of their cars. Yet we should not credit the AMA resolution with all that was wrong with American cars in 1958. The normal styling lead time meant that the lines

Zora Arkus-Duntov (left) with road-tester Tom McCahill, looking over an early 1957 Corvette with optional factory hardtop. The fender scoops were dummies.

McCahill and Duntov view the fuel-injected 283 cid engine on the 1957 Corvette. Developed by Rochester to a GM design, the 'fuelie' greatly improved performance, but was unreliable on the circuits. Only 240 of the 1957 models were so equipped.

for the bulbous monstrosities fielded by GM and Ford that year were laid down in 1955, when everyone was racing their hearts out. So it was a non-vintage year from several standpoints — and it was not the greatest year for the Corvette.

Compared to the '57, the '58 model was heavier and much busier looking, especially at the front end. Following a fad, the '58 adopted quad headlamps, and from them a heavy piece of bright metal trailed back along the tops of the front wings. The bonnet sprouted simulated louvres, and more chrome appeared on the side and rear end. The body of a formerly trim roadster was 10in longer and 3in wider than the year before — and at 3,000lb it weighed a good 100 more at the kerb.

Some of the positive changes were evident inside the car and under the bonnet. At Zora Duntov's behest the dashboard was redesigned, grouping the instruments under the steering wheel and placing the rev-counter where it should have been in the first place. The fuel injection was said to be more reliable, and made the engine somewhat more powerful — 250 and 290bhp were its ratings — while normally aspirated engines offered 230/245/250/290bhp. RPO 684 being still on the books, Corvettes won the Class B-Production national championship of the Sports Car Club of America in 1958 and 1959 — without any overt help from GM, of course.

The heavier styling was mainly the legacy of Harley Earl, who

A heavier, busier look was in vogue for 1958, when Corvette joined the quad-headlamp generation. The body was 3in wider and 10in longer and the car weighed about 100lb more than the 1957 version.

A 1958 model, apparently a showcar (note the narrow-band whitewall tyres). Another change for 1958 was the reverse scoop on the front wings, filled with bright metal 'teeth'.

This production model 1958 Corvette hardtop shows that the lithe lines of the 1956-57 models were to some extent squandered this year. Bill Mitchell didn't much like the style and when Harley Earl retired he set himself to altering it.

Flush-mounted tail lamps were a functional feature of the 1958 rear design, but the large chrome bands and the erotic bumpers were not.

retired (in favour of Bill Mitchell) in 1958, and whose designs Mitchell wasn't able to entirely replace until about 1961. Following the usual Detroit cycle, the 1959 and 1960 models were pretty much the same, though Duntov and other enthusiasts managed to get those fake louvres removed from the bonnet. There was a lock-out system on the shift lever, and the clutch had a wide range of adjustment. A new option was RPO 686, sintered metallic brake linings, which cost the buyer merely $27 and were well worth the money in increased resistance to fade. To reduce axle tramp, radius rods were fitted to the rear suspension.

Mitchell was working on a proposed 1960 Corvette called the Q-car, with independent rear suspension and a rear transaxle on only a 94in wheelbase. The design closely presaged that of the 1963 Sting Ray. But these were not particularly good years at Chevrolet, where radically finned passenger cars were losing many sales to Ford, and the Q-car was eventually shelved in favour of a little-changed 1960 Corvette.

The '60 models did see more use of aluminium, in clutch housings, radiators and fuel injection cylinder heads, but unfortunately the latter tended to warp if the engine overheated, and aluminium heads were quickly dropped. Unique (a first on an

Zora Duntov helped convince designers to improve the 1958 cockpit by putting the rev-counter directly in front of the driver, flanked by auxiliary gauges. The console contained the radio and heater controls and an electric clock.

American car) was a rear suspension anti-sway bar, which replaced the stiff-spring arrangement and gave a better ride with no loss in roadability. And 1960 was also the year that Corvette enjoyed its finest hour, at the 24 Hours of Le Mans.

Briggs Cunningham had entered three cars in the GT class that year. One of them, driven by Bob Grossman and John Fitch, was soon putting in 151mph sprints down the Mulsanne Straight, and this Corvette eventually finished eighth in a potent field. It was a major success for the marque, and another indication that if GM was officially out of racing, there were plenty of private individuals who weren't. Happily the spares remained on the shelves to 'prodify' Corvettes into competitive world-class racing cars.

For 1961, new Chief of GM Design Bill Mitchell was able to work some modification to the Chevrolet sports car, chiefly at the rear end. Basing his ideas on his own racing Sting Ray and XP-700 showcar, Mitchell gave the '61 a clean, reverse-canted tail with inset tail-lamps, and threw out the grille teeth that had

Fake louvres were removed from the bonnet of the 1959 Corvette models, which otherwise looked pretty much the same. Sintered metallic brake linings were a new option; they cost only $27 and much improved fade-resistance.

The lean lines of the 1959 Corvette are emphasized in this side view of an open model, as is the angle of the support pillars for the wraparound screen.

Three-quarter rear view of a similar car revealing a much cleaner rear-end style compared with the 1958 hardtop model pictured on page 35.

The 1960 Corvette roadster with fuel injection. With 4.11:1 rear axle ratio, this car was capable of standing quarter-mile times in 14 seconds at up to 100mph.

distinguished (or denigrated) Corvettes since the beginning. Instead, there was merely a mesh screen. Headlamp rims were now painted the body colour, further helping to present a cleaner look, and standard equipment included an aluminium radiator, parking brake warning lamp, dual sun visors, interior lights and windscreen washers. Engine options were unchanged in 1961, but the four-speed gearbox received an aluminium case and the three-speed a wider choice of gear ratios. Automatic transmission was still available, but only 15 per cent of Corvettes were so equipped.

The top-line 1960-61 engine was the 315bhp version with fuel injection. Despite an increase of 20bhp over the 1959 290bhp 'fuelie', performance wasn't much affected. All you could do with a 315 fuel-injected Corvette with 4.1:1 axle was a 100mph standing quarter-mile in about 14 seconds. Same old stuff!

For the 1962 model year Bill Mitchell further refined styling of

Outwardly little changed, the 1960 saw wider use of aluminium (including clutch housing, radiator and fuel injection cylinder heads). On the 'performance' suspension option, a rear anti-sway bar was used instead of stiffer leaf springs. At Le Mans, Briggs Cunningham's Corvettes were timed at 151mph on the Mulsanne straight.

For 1961, Bill Mitchell's influence showed up at both front and rear. The Corvette lost its toothy look as Mitchell replaced the vertical grille bars with a neater grid; headlamp rims were painted body colour and front wing strips vanished. In the rear, a ducktail design appeared from the racing Stringray and the XP-700 showcar, with inset tail lamps and nerf bars.

Cleanest of the four-eyed Corvettes was the '62, with its blacked-out centre grille. By comparison with more modern designs like the Jaguar E-type, however, the Corvette was clearly ageing, and new styling was on the way for 1963.

aluminium was added to the rocker panels. Stiff springs came back to the option list in 1962, and with their help Dr Dick Thompson won the SCCA Class A-Production national championship for Corvette. (The hairier cars had been bumped up from Class B, where they had been shattering their rivals — SCCA freely moves cars around to classes where they are thought to be competitive, without much regard to their displacement.)

Also in 1962, Chevrolet Division received a new General

The cockpit of a 1962 Corvette roadster. Note the considerable extra bolstering of the seats and the different door trim compared with the 1958 model illustrated on page 36.

the now quite aged Corvette body; it was five years old and due to be swept away forever in 1963. Mitchell eliminated the bright metal outline around the concave side moulding, and replaced the three long 'teeth' inside the reverse-mounted side scoop with a grid. The mesh grille was blacked out and a strip of anodized

Mitchell cleaned up the sides of the '62 Corvette by deleting the bright metal outline around the concave section and replacing the reverse scoop's teeth with a fine row of tiny blades. The anodized aluminium rocker panel was designed to enhance the low look.

Close-up of the decorative blades which define the leading edge of the concave part of the body side of a 1962 Corvette.

Manager. As Ed Cole left for higher duties, Semon E. 'Bunkie' Knudsen took over and determined to get Corvette production up to 15,000. He nearly succeeded; 14,531 of the '62 models were built, compared to only 10,939 of the '61s. Corvette had been profitable to GM since 1958 — Knudsen now saw that it was paying a really good return on investment.

Another important 1962 milestone was the new 327 cubic inch V8, produced by enlarging the 283 to a 4.00in bore and 3.25in stroke. Fuel injection was modified to suit the new engine, which would become a long-running Corvette standby. A high-speed 'cruising' rear axle ratio of 3.08:1 was offered for the two lowest-horsepower engines. The horsepower range of the 327 in 1962 (and 1963) was 250 to 340bhp with carburation and 360bhp with fuel injection. The 327 was an all-round better engine than the 283, which hadn't been bad. Mid-range performance was up, and those 15-second, 100mph quarter-mile times formerly available only with 4.11 axles could now be achieved by a carburettor 327 with a 3.70:1 rear axle ratio.

The second great era in Corvette development, which began with the restyled '56, had now reached its end and its peak.

This is the corporate emblem which was featured on the rear deck of a '62 Corvette, where it was sited on top of the central rib which ran rearwards from the seat divider at the back of the cockpit.

A somewhat simpler motif, incorporating the crossed flags, but without the superimposed Vee, was displayed at the front of the car on the panel linking the engine cover and the grille.

Thanks mainly to Zora Arkus-Duntov, the Corvette had made important progress through these years, despite the heavier models brought in for 1958 and the AMA 'ban' on racing. All traces of the original 1953-55 cars were gone; the cars were quicker than ever, better handlers than they'd been before, and more cleanly styled for 1962 than in any year since 1957.

Among collectors, there is a clear preference for model year 1956, 1957 (especially with fuel-injected V8s) and 1961-62 Corvettes. The Milestone Car Society, which recognizes 'the great cars of 1945-67', has granted Milestone status to all Corvettes through to 1958, the 1960 and 1962-67 models, but curiously leaves out the '59s and the '61s. Why they should not recognize the '59 if they accept the '58 is not clear. But it is true that of all the model years in this period, 1958-59 was a low point. If only through deft facelifts, the cars improved steadily in 1960, 1961 and 1962. Frankly, for this writer's money, the '62 was the highest developed and most refined of the lot, and therefore the best one to have.

CHAPTER 4

Early Corvettes in competition

1953 to 1962

As hard as it may be for you to believe, some people did race the early two-seat Thunderbirds, but nobody has recently admitted racing the early Corvettes — at least there's nobody for modern writers to talk to about it. If any of the 1953-55 models did appear on SCCA road courses, they were roundly trounced and did not show up often. Officially, the 1953-55 Corvette was relegated to SCCA's Class C-Production. This class then contained the Mercedes-Benz 300SL Gullwing, which weighed about the same but had 100 more bhp, better handling, better brakes and a proper four-speed gearbox. Enough said.

In 1956, however, things changed in a hurry. Zora Arkus-Duntov had reworked the Corvette chassis, relocating the rear springs and resetting the front suspension. The new V8 engine was at hand (as it had been in 1955, though the '55s didn't race much) and to it Duntov applied a special camshaft of his own design. The result was a 250bhp Corvette, which scored 150.583mph at Daytona in February 1956 — driven by Duntov himself. Two other similar cars were driven to 145 and 137mph by John Fitch and Betty Skelton, respectively.

For the 1956 12 Hours of Sebring, John Fitch prepared a team of four Corvettes, using the Duntov cam, ported manifolds and two four-barrel carburettors. One car was bored to the 5-litre limit (307 cid) and equipped with a ZF four-speed gearbox for the prototype class. All four cars used Halibrand magnesium wheels, driving lights and oversize fuel tanks to minimize pit stops. A car driven by Walt Hansgen and John Fitch finished ninth overall, and the next Corvette came 15th, with Ray Crawford and Max Goldman up. It was a good showing for the first time out.

In mid-1956 Duntov created another special with Sebring specs and an extended aerodynamic nose and a small tail-fin on the boot lid. He called it the SR-2, and it was raced extensively by Dick Thompson and Curtis Turner. In 1958, under Jim Jeffords, SR-2 was still racing — and it went on to win the SCCA Class B championship, painted a bilious purple and bearing the legend, 'Purple People Eater'. Another SR-2, campaigned by Bill Mitchell and driven by Pete Lovely, had finished 16th at Sebring in 1957.

The well-known Washington DC dentist, Dr Dick Thompson, began driving Corvettes as soon as they were raceworthy. In 1956, Duntov prepared a C-Production roadster for Thompson. Thompson's success helped convince General Motors that the Corvette had a competitive future, and thus brought about the fuel injection and four-speed gearbox options on the 1957 models. Also in 1957, RPO 684 brought the competition suspension mentioned in the previous chapter — an all-out racing set-up right out of the option book. All these factors, plus the enlarged 283 V8, bumped the Corvette into SCCA's B-Production class for 1957. This didn't deter Thompson, who won his second straight SCCA championship that year. Thompson also entered the GT class at Sebring, teaming with Gaston Andrey. The pair won their class handily and romped home 12th overall.

In July 1956, Duntov began work on a new sports-racing car called the Sebring SS, using a tubular spaceframe, a fuel-injected 283 V8, a magnesium body and a De Dion rear axle. But on its debut at the 12 Hours, in 1957, the SS retired after just 23 laps. Duntov's plan to fight again at Le Mans was then thwarted by the Auto Manufacturers Association's anti-competition resolution.

Though the Sebring SS died stillborn as a racing car, it did lead

Sebring 1958. 'Faster than a speeding bullet', a Corvette gets the jump on the pack at the Le Mans start. Jim Rathmann and Dick Doane shared the GT class winner. *(Corvette News)*

Typical view of a winner. Norm Munson takes the overall and C-Production checker at a Marlboro, Maryland Regional SCCA race in 1956 after vanquishing Charlie Wallace's Mercedes 300SL challenge. *(Tax Rufty)*

Corvettes in action in 1956. Top left, Bark Henry pressing Charlie Wallace's Mercedes 300SL at Cumberland, Maryland. *(Dave Wheeler)* Top right, Dick Thompson pierces the fog at the top of Mount Washington, New Hampshire, in establishing the fastest production car time in that year's hill-climb. *(Fred Vytal)* Above left, Fred Windridge dices with the Jaguar XK120 of Jim Jeffords in an SCCA race at Elkhart Lake, Wisconsin. *(Barney Clark)* Above right, a pair of Corvettes neck and neck in the Elkhart Nationals. *(Barney Clark)*

The 1957 model makes its debut at the Nassau Speed Weeks. Much experimentation on fuel injection and suspension settings was done here for the upcoming Sebring 12 Hours. *(Barney Clark)*

to another, better competition Corvette — the Stingray. This smooth, aggressive-looking car was designed by Bill Mitchell on the chassis of the SS 'practice mule', and was handed over to Dick Thompson for C-Sports Racing on the SCCA circuits during 1959 and 1960. In 1960 the Stingray won its class title — in spite of the fact that Thompson never won a race outright! Consistent high placings had given him the championship on total points.

Recalling the Stingray, Dick Thompson said it was almost a perfect car — except that it wouldn't stop: 'It got pretty hairy at times, and I had a few accidents because of those brakes'. (These included flipping the Stingray at 100mph at Meadowdale, landing upside-down.) 'Once at Laguna Seca, in the last turn, Walt Hansgen passed me, drove in front, then put his brakes on. I had no brakes at all, there was no escape road in those days, there were people all around and I had no place to go — so I hit him. When I say I had no brakes, I mean *none at all*. I didn't even bother putting the pedal down. . . . Fortunately, the car handled well, so when I ran out of stop I could slide it sideways and take off a good bit of speed that way. I had to "dirt track it" around. I always made sure it had a little bit of oversteer, so I could get the tail end around.'

More important, perhaps, from the standpoint of history, the Stingray influenced production designs. Both its name and its overall lines were to appear on an all-new line of Corvettes launched for 1963 — the first designed entirely under Mitchell. The colourful designer told the writer: 'That racing thing was just camouflage — we wanted to get visibility for the car, help get that styling into production'. Yet the Stingray was a better racing car than many all-out competition rivals. Indeed it was all things to all men — a championship racer, a sexy showcar and a novel prototype. It's one of the most important Corvettes.

Corvette collectors should know that this influential sports car still exists — in the hands of GM Styling's Special Vehicles Section. In 1962, it was 'prodified' with a passenger seat and full windscreen. Bill Mitchell used it for personal transportation — and just to remind himself about traffic police he even installed a speedometer. Later it was fitted with Dunlop disc brakes — Dick Thompson said these corrected its greatest fault — a prototype 377 cid V8 and, finally, the big 427 powerplant. The latter engine breathed through four Weber carbs which poked through the bonnet, and in this form the Stingray was painted bright red. Ultimately Mitchell repainted the car silver, refinished the interior, smoothed off the bonnet and reinstalled Duntov's special 377 engine. Eschebach and Henderson brought the car up to concours condition for the New York Auto Show in 1977 — and thus it remains today.

During the late-'Fifties, Corvettes were virtually unchallenged in big-bore Production and Modified racing. Even the E-type Jaguar, which arrived in 1961 and was soon tried on the road courses, was no match for it. Jim Rathmann and Dick Doane won the Sebring GT class in 1958, then Jim Jeffords followed up his 'Purple People Eater' title that year with another B-Production class championship in 1959. In 1960, while Dick Thompson was successfully campaigning the Stingray in the Modified class, Bob Johnson was winning B-Production and Chuck Hall/Bill Fritts took the GT class at Sebring. And 1960 was also the year when one of Briggs Cunningham's cars was placed eighth overall at Le Mans — this Corvette was driven by John Fitch and Bob Grossman.

Dick Thompson, switching back to B-Production for the 1961 SCCA racing season, won his class going away. Don Yenko repeated for Corvette in 1962 and 1963, when Thompson won this class and then retired. Finally, in 1964, Frank Dominiana took the B-Production title — but this was the last SCCA championship for the Corvette for many years. Another finale, as it proved, was the GT class win at Sebring in 1961 with Delmo Johnson and Dale Morgan up. After that, a long dry spell set in.

What happened had nothing directly to do with the Chevrolet Corvette, or even the notorious AMA 'ban' on racing. Carroll Shelby had entered the picture by dropping a small-block Ford V8 into an AC Ace and producing the first Cobra. 'It was clear as day to me', recalled Zora Arkus-Duntov, 'that the Cobra had to beat the Corvette. [It] was very powerful and weighed less than 2,000 pounds. Shelby had the configuration, which was no damn good to *sell* to people, except a very few. But it had to beat the Corvette on the tracks.' Duntov's contention is easily proven by the statistics. Total production of Cobra 289s over the 1962-65 period was 665 (including 75 260 cid V8s, 24 competition roadsters and six coupes). Total production of Corvettes over the same period was 81,835!

The pits at Sebring in 1957. Dick Thompson and Gaston Andrey eventually won the GT class and finished 12th overall, completing 173 laps against the Behra-Fangio Maserati's 197. *(Dave Wheeler)*

SPORTS CAR CLUB OF AMERICA CHAMPIONSHIPS 1956-62

C-Production
1956: Dr Richard Thompson, Washington DC

The experimental Sebring SS, under discussion by driver John Fitch (left) and Zora Arkus-Duntov at Sebring in 1957. The car had completed only 23 laps before a spate of nits caused its retirement. A rubber bushing at the chassis end of one rear trailing arm had shifted out of place, and there was severe tyre scarring caused by locking front brakes. *(Motor Racing)*

Fitch and the SS corner at Sebring (where cornering was not something one had to do particularly well). Before problems set in, the SS was competitive, clocking laps around 3 minutes 32 seconds and turning several in at below 3:30. But more development work was needed. *(GM Photographic)*

The SS fitted with a plastic bubble top. It never raced again after Sebring 1957. *(Henry Ford Museum)*

Ned Yarter of La Crescenta beats the Jaguar of Bob Kudler by a few car's lengths during a Triple-R sports car road race for production cars at Bakersfield, California, in June 1957. *(Motor Racing)*

B-Production
1957: Dr Richard Thompson, Washington DC
1958: James Jeffords, Milwaukee, Wisconsin
1959: James Jeffords, Milwaukee, Wisconsin
1960: Robert Johnson, Columbus, Ohio
1961: Dr Richard Thompson, Washington DC
1962: Donald Yenko, Canonsburg, Pennsylvania

A-Production
1962: Dr Richard Thompson, Washington DC

C-Sports Racing
1960: Dr Richard Thompson, Washington DC

B-Sports Racing
1957: J. E. Rose, Houston, Texas

STINGRAY PROTOTYPE RACING RESULTS

1959
Apr: 1st in class, 4th o/a, SCCA National, Marlboro, Maryland (R. Thompson)
Jun: DNF, SCCA National, Elkhart Lake, Wisconsin (J. Fitch)
Jul: 9th o/a, USAC Professional Series, Lime Rock, Connecticut (R. Thompson)

Despite the rain-slick track, Bert Ruttman's Corvette dominated the over-2,700cc production car race at Pomona, California, in February 1958. His car was a 1957 'fuelie'. The AMA 'ban' was now in effect, and Corvette campaigning was strictly by privateers. *(Motor Racing)*

A 1962 Corvette in the pits at Road America, Elkhart Lake, Wisconsin. This would be a lower-horsepower, non-injected car in Class B-Production. By this year, the Corvettes were fast being outclassed by the Cobras. *(Goodyear Tire & Rubber Co.)*

Jul: DNF, USAC Professional Series, Meadowdale, Illinois (R. Thompson)
Aug: DNF, SCCA 500-mile Endurance Race, Elkhart Lake, Wisconsin (R. Thompson)
Dec: 10th o/a, Trophy Race, Nassau, Bahamas (R. Thompson/ A. Lapine)

1960
Apr: 1st in class, 3rd o/a, SCCA National, Marlboro, Maryland (R. Thompson)
May: 1st in class, 2nd o/a, SCCA National, Danville, Virginia (R. Thompson)
May: 1st in class, 5th o/a, SCCA National, Cumberland, Maryland (R. Thompson)
Jun: 2nd in class, 2nd o/a, SCCA National, Elkhart Lake, Wisconsin (R. Thompson)
Jul: DNF, SCCA National, Meadowdale, Illinois (R. Thompson)
Sep: 2nd in class, 23rd o/a, SCCA 500-Mile Enduro, Elkhart Lake, Wisconsin (R. Thompson)
Sep: DNF, SCCA National, Watkins Glen, New York (R. Thompson)
Oct: 11th o/a, SCCA Professional Series, Riverside, California (R. Thompson)
Oct: DNF, SCCA Professional Series, Laguna Seca, Californi (R. Thompson)

CHAPTER 5

The Sting Ray era

1963 to 1967

Among collectors, no generation of Corvettes is admired as much today as the 1963-67s — the first production Sting Rays. The 1953-55s are perhaps more costly, due to their extremely low production; the 1956-62s have a far more impressive competition history; the 1968-82s are more numerous and their body design lasted a lot longer. But for Corvette enthusiasts the Sting Ray generation remains special — and in terms of value appreciation, these are the best Corvette investments you can make today.

The 1963 Corvette Sting Ray represented the most radical design and engineering change in the history of the marque. Except for the established range of V8 engines, almost everything about the car was different from the year before. Styling was completely new, and for the first time a coupe body style joined the open model. The coupe contributed over 10,000 sales, moving Corvette production over 21,000 for the 1963 model year — some 50 per cent better than any previous year on record.

Designs for the new-generation '63 began in 1960. The chief goals of the project, according to Duntov, were better driver and passenger accommodation, better luggage space, better ride, better handling and higher performance. As a styling model, Bill Mitchell's GM Styling Staff proposed the XP-720 project car, which had been finished in late-1959. The XP-720 was based on the lines of the Stingray racing car, but featured its own unique fastback coupe styling above the beltline and a distinctive divided rear window. The divided backlight was Bill Mitchell's personal idea, and getting the more practical Duntov to accept it wasn't easy. In production, it appeared only on the 1963 model. Mitchell thought it was an important component in the overall look — 'Take that off and you might as well forget the whole thing', he said later.

(Not to be pedantic, but we should mention here a spelling difference for the purpose of accuracy. The original racing car was dubbed 'Stingray', while the production Corvettes of 1963 through to 1968 used the more proper two-word designation 'Sting Ray'. For reasons known only to the Sales and Marketing Departments, Corvette went back to 'Stingray' with the restyled 1969 production models. The name was at least ostensibly retained into the 'Eighties, with less and less emphasis by GM.)

It should be remembered that the major thrust of Sting Ray design came in late-1960 and 1961 — at precisely the time the E-type Jaguar was wowing sports car enthusiasts worldwide with one of the most timelessly beautiful shapes ever to adorn a production automobile. Bill Mitchell was a strong admirer of the E-type — he owned and drove one personally, to the consternation of his superiors — and much of the '63 Corvette's functional beauty was encouraged by the flawless lines of Sir William Lyons' masterpiece. Still, there were some unfunctional aspects to the Sting Ray; dummy scoops in the front quarter-panels, pseudo air vents in the bonnet, and Mitchell's two-piece backlight.

An early alteration from the design of the Stingray racing car was the switch to hidden headlamps, achieved via pivoting leading body sections, mounted flush in the sharply creased front end. Styling Staff built a 'dip' into the beltline at the trailing edge of the door, and on coupe models the doors cut into the roof to provide ease of entry and exit. Some of these ideas harked back a long way: the ill-fated 1948 Tucker originated the idea of cut-in doors, and hidden headlamps had been around since Gordon

All new from the ground up, the 1963 range included the first Corvette coupe. Styling was based on the racing Stingray and XP-720 prototype. Bill Mitchell's designers achieved an air of aggressive function which looked fast standing still — and faster in motion.

Wind-tunnel testing of a quarter-scale clay model at Cal Tech. Mitchell and Duntov agreed that the production car must be as slippery as possible.

Buehrig's classic prewar Cords, though the Corvette was the first production American car to use them since the 1942 DeSoto, of all things. Another DeSoto — the 1955-56 series — featured a 'gullwing' dashboard, though it's doubtful that it was the actual precursor of the Corvette's: Mitchell said the new dash was 'essential to the whole package. It was widely criticized at the time, but it was a very fresh approach to two-passenger styling, and I think it worked remarkably well'.

Several high-ranking Chevrolet and GM executives had looked with envy on Ford's success with the four-seater Thunderbirds and had lobbied for a similar Corvette. Unlike Ford, Chevy did not plan to replace the two-seater, merely to expand the line. Ed Cole himself was among four-seater advocates, but both Bill Mitchell and Zora Duntov opposed this. They said a four-seater had destroyed the Thunderbird's sports car pretensions and that the Corvette had always been and should remain a genuine sports car. The opposition then suggested a two-plus-two (which Jaguar later tried, without much aesthetic success). But when Mitchell and Duntov stood together, there was no-one who could gainsay them. The Corvette coupe was as far as they'd go towards expanding the line — and this model was strictly for two. The coupe proved a tremendous success, and the '63 actually leads the open cars in collector appeal today — which is an uncommon situation.

Prototype models of the Sting Ray coupe and roadster were subjected to intensive evaluation, including wind-tunnel tests at the California Institute of Technology. GM body engineers took pains to redesign the inner body structure accordingly. For instance, the 1963 Sting Ray had almost twice as much steel support built into the central body structure as the previous models — but the added weight was compensated by a reduction in the amount of glass-fibre body panels, and the '63 actually weighed slightly less than the '62.

One reason the '63 was lighter was that its wheelbase had been reduced by 4in. Its rear track was 2in narrower and frontal area (in the interests of aerodynamics) was reduced by 1sq ft. Yet interior space was at least as good as before and — thanks to the extra steel reinforcement — the cockpit was stronger and safer.

Other body features included curved side windows, cowl-top ventilation, more luggage space and an improved fresh-air heater. No lid was designed into the rear of the coupe, however, and one

The divided backlight was a major bone of contention at Chevrolet Division. Zora Arkus-Duntov felt visibility was affected, but Mitchell said, 'Leave that out and you might as well forget the whole thing'. Doors cut into the roof were a unique feature of the coupe.

Body drop at the St Louis assembly line. Though the 1963 glass-fibre body had added steel reinforcement, it was commensurately lighter than the 1962 version, resulting in a slight loss of overall weight.

Final inspection at St Louis. Coupe and convertible models each sold more than 10,000 units for the 1963 model year, giving a higher total production than in any previous year. The quality of fit and finish was considerably higher than the GM standard of the day.

Office of the '63. Still a bit too much glitter here for serious enthusiasts, and the parking brake was a clumsy pedal-and-pull-handle affair. Mitchell designed the 'gullwing' dashboard to individualize driver/passenger compartments. The rev-counter read to 7,000rpm, the speedometer to 160mph.

had to reach behind the seats to add or remove luggage. The spare tyre, housed in a special holder which dropped to ground level for access, was easier to get at than the luggage compartment.

While the engine line-up was unchanged from 1962, the rear suspension was redesigned, becoming a fully independent three-link type with double-jointed open drive shafts, control arms and trailing radius rods at each side. A single transverse leaf spring was mounted to the frame with rubber-cushioned struts — the leaf was used because the body permitted insufficient room for coils. Duntov bolted the differential to the rear crossmember, insulating it with rubber at the mounting points. The frame itself was a well-reinforced box-type. Weight distribution was now 48/52 per cent (as opposed to 53/47 on the 1962 model), and both ride and handling were improved. Axle tramp on hard acceleration was completely eliminated.

Steering was a new recirculating-ball type combined with a

A 327 V8 installation in a Sting Ray. Available throughout the 1963-67 generation, the 327 offered flexibility with a horsepower range from 250 to 365. The last fuel-injected versions were made in 1965.

An all-new link-type rear suspension with transverse leaf spring replaced the conventional beam axle for 1963 and later models. Control arms and radius rods were used. The body configuration mandated a single leaf spring rather than coils.

Wide-angle lenses will do interesting things to Sting Ray coupes, in this case emphasizing Mitchell's racy rear-end styling and his divided backlight theme on a '63.

dual-arm, three-link ball-joint front suspension. Fewer turns lock-to-lock were required than before. Drum brakes were still used all round, but the front drums were wider and all brakes were self-adjusting. An alternator replaced a generator; positive crankcase ventilation made its first appearance; the flywheel was smaller and lighter; and the clutch housing was a new aluminium unit.

By 1963 the big manufacturers were paying lip-service (at best) to the 1957 anti-racing resolution, and competition options were abundant. These included heavy-duty springs and shocks, a stiffer anti-sway bar, metallic brake linings, optional AlFin aluminium brake drums, cast aluminium knock-off wheels, a dual master cylinder, and a long-distance fuel tank holding 36.5 US gallons (29.2 Imperial). Full leather upholstery became an option for the first time, giving the cars a useful modicum of added luxury.

Road test magazines were generally enthusiastic about the new Sting Ray, admiring its styling and raving about its improved handling. Axle tramp and serious oversteer, ever characteristics of the pre-1963 generations, had been eliminated. 'Every time through [our slalom course] we discovered we could have gone a little faster', noted *Road & Track*. 'We never did find the limit.'

Contrary to the traditional American practice of adding trim each new model year, Chevrolet actually subtracted it. For the 1964 model year, the split backlight and fake bonnet louvres were dropped. Slotted wheel discs were added to improve brake ventilation. The extractor vents on the coupe were redesigned and made functional for the first time. GM policy was to make improvements in an evolutionary sense; wisely they realized that a design so good in the first place needed no severe alterations to meet the annual new car model hoopla. Thus the Corvette did not suffer from clutter in these years, when 'progress' to most Detroit manufacturers was symbolized by extra tail-lights, new expanses of brightwork or glittery new medallions. Thus the Corvette improved its international reputation as a serious sports car rather than a plastic plaything.

In 1965 another piece of non-function vanished when the slots in the front quarter-panels were opened, allowing them to duct heat out of the engine compartment and away from the front wheels. The extractor vents of the coupe, which had been opened

Front and rear views of the 1963 coupe. Ornamentation was a bit overdone, but it gradually decreased in later years. Bonnet grids were fakes, but front-end sections pivoted to expose headlamps. An oversize fuel filler was contained under the rear medallion.

The convertible for 1963 with the 360bhp fuel-injected 327 V8. Injection models were continued only until the big-block Mark IV engine arrived in early 1965. They failed to sell well and proved troublesome in service.

The 1963 Corvette convertible, with and without the optional factory hardtop.

There's something in what Mitchell said. Dropping the divided coupe backlight for 1964 did take away some of the character of the original coupe — while providing better visibility. Collectors have shown a marked preference for the split-window version, whose current market value exceeds that of Corvette roadsters.

up in 1964, had proven inefficient, so they were eliminated altogether for 1966. An egg-crate-style grille was also adopted that year. In 1967, the only changes were an oblong reversing lamp, revised front fender louvres, bolt-on instead of knock-off aluminium wheels and an optional black vinyl covering for the roadster's removable hardtop. The '67 was thus the cleanest version of the Sting Ray generation — a beautifully finished car which still looks elegant today.

Mechanical improvements were rightly stressed over styling gimmicks in these years, and each new Corvette model brought its share of them along. The fuel-injection 327 cid V8 of 1963 developed 375bhp (1.15bhp per cubic inch), and delivered 100mph from rest in only 15 seconds. For 1965, disc brakes finally arrived — not just up front, but on all four wheels. This cured a long-standing fault with the car and the elimination of drum brakes was well praised. A Corvette needed to stop as well as go, for engines were getting increasingly powerful — as typified by 1965's new Mark IV V8 with 425 horsepower.

The Mark IV was not the first big-engined 'Vette. Micky Thompson had built specials for Daytona with the 409 cid Chevrolet Super Sport passenger car engine in 1962; Zora Duntov himself had tried an experimental 377 in the Stingray racing car. But Duntov generally resisted the idea of a production Corvette with a huge engine. What changed his mind was

The 1964 Sting Ray coupe. Changes this year included slotted wheel covers for improved brake cooling and redesigned extractor vents in the roof pillar, which were made functional for the first time.

The 1964 Sting Ray convertible, cleaner than the 1963 through lack of bonnet grids. Otherwise, the car was much the same as in the previous year.

Shelby's Cobra, which by 1964 was overwhelming Corvettes on the nation's road courses. The Mark IV was engineered for the Corvette by Duntov and Jim Premo, who had replaced Harry Barr as Chief Engineer of Chevrolet.

The initial plan was for a 396 cid V8 (bore and stroke 4.09 x 3.76in), because GM corporate policy banned the use of engines over 400 cid on anything except the full-size Chevrolet passenger cars. An exception was made for the Corvette in 1966, when the 396 was increased to 427 cubic inches. In its 1965 396 cid form it replaced the 365bhp small-block engine.

This was a formidable performance powerplant with four-barrel carburettor, solid lifters and 11:1 compression ratio. Its 425bhp was accompanied by 415lb/ft of torque at 4,000rpm. To handle this brute force, the 396 Corvettes came with stiffer front springs, a thicker front anti-sway bar, a new rear sway bar, a heavier clutch and an oversize radiator and fan. Though the 396

The instrument and control layout of a 1964 Corvette. Dials were neatly grouped ahead of a dished steering wheel and the short gear lever was well placed on the centre console.

The emblem on the rear deck of the 1964 Corvette emphasized the use of two words to record 'Sting Ray' (though without any space between them in this instance) as distinct from the 'Stingray' identification used for the earlier Corvette racer and the later production series.

Body painting a 1966 coupe with the big Mark IV V8 engine scheduled to be installed under the bulging bonnet. The Mark IV arrived in 1965 with 396 cubic inches, but it was bored to 427 cid in 1966; a 'cooking' version with 4.11:1 axle could deliver 0-60mph in less than 5 seconds.

Mark IV engine weighed over 650lb, weight distribution remained near-neutral at 51/49 per cent. The cars were easily distinguishable by a large bulge in the bonnet and were usually equipped with the optional side-mounted exhaust pipes.

The 427 of 1966 was achieved by increasing the bore to 4.25in. This produced the fastest street-stock Corvette in history. With the 4.11:1 rear axle ratio, *Sports Car Graphic* recorded 0-60mph in an astonishing 4.8 seconds, 0-100mph in 11.2 seconds, *and* a flat-out maximum speed of 140mph. The 427 (Ford-engined) Cobra could generally just exceed these figures — but as Duntov was wont to point out, the Cobra was more a competition car than a Grand Tourer. Despite its power, the 427 Corvette remained a comfortable high-speed GT which was remarkably tractable, even in stop-and-go traffic.

A casualty of the Mark IV engine programme was fuel injection, which had never come up to Chevrolet's expectations and was dropped from the option list after 1965. To sort out the plethora of engine options in the Sting Ray period, we'd do well

The 1965 Sting Ray coupe with fuel injection — one of the last such installations. Slots in the front quarter-panels had now been opened, allowing them to duct away engine heat.

to list them here:

327 V8
250bhp @ 4,400rpm, 1963-65
300bhp @ 5,000rpm, 1963-67

340bhp @ 6,000rpm, 1963 only

Mark IV V8
390bhp @ 5,400rpm, 1966-67
400bhp @ 5,400rpm, 1967 only

425bhp @ 6,400rpm, 1965 only**

350bhp @ 5,800rpm, 1965-67
360bhp @ 6,000rpm, 1963 only*
365bhp @ 6,200rpm, 1964-65*
*fuel injection

425bhp @ 6,400rpm, 1966-67
435bhp @ 5,800rpm, 1967 only
**396 cid (others all 427 cid)

In 1965, Bunkie Knudsen stepped down as Chevrolet General

The 1965 convertible saw few changes once again, although disc brakes were new this year.

Manager in favour of Elliott M. 'Pete' Estes, later President of GM. Knudsen's tenure at Chevrolet had clearly established Corvette as a high-profit segment of the Chevy line-up, and in doing so had guaranteed its permanence as 'strictly a sports car'. Unlike the Thunderbird, the Corvette would never be watered down into something unrelated to its original concept. With sales of over 20,000 units a year, it had become a respected and permanent part of the Chevrolet car line-up, and better sales years were still to come.

Collectors have often remarked on an irony — that the 1963-67 generation was the best Corvette series ever, yet it was the shortest-lived series since the original 1953-55 line. But this was inevitable. Bill Mitchell says, 'we were working on a successor about two months after the '63 had appeared in the showrooms'. At that time GM was wedded to a three or four-year styling cycle, and even when the all-new '68 was introduced plans were afoot for yet another new Corvette generation for 1971 or 1972. (What prevented those plans from realization is the subject of another chapter.)

The shortness of its tenure does not prevent us from labelling the 1963-67 Sting Ray as the definitive Corvette — the finest evolution of the American sports car as visualized by Zora Arkus-Duntov, Bunkie Knudsen, Ed Cole and Bill Mitchell — car lovers all. It is no surprise that these four years' worth of Corvettes sold nearly 120,000 units and that they remain, as a group, the most sought-after Corvettes today. They brought the marque to the pinnacle of development — beautifully engineered, gracefully proportioned, blindingly fast on all types of roads. It is no surprise that all four vintages are on the list of Milestone cars.

For collectors today, there's a strong dichotomy of opinion about Sting Rays. One body of enthusiasts will opt for the coupe models (which were always a minority of production), another body for the open cars. These groups then subdivide into supporters of the small-block 327 engine and of the Mark IV big-incher. So there are really at least four different ways to approach the Sting Ray from a collector's standpoint.

For this writer's money, however, the small-block cars are the way to go, at least if you're interested in all-purpose motoring — touring as well as competition. Particularly in North America, the big-block cars are increasingly hard to live with. Our petrol has a maximum octane rating of 94, unless you buy it at an airport (and

The 1966 Sting Ray coupe. The roof pillar extractor vents, having proven inefficient, were eliminated with this model.

some Mark IV owners do). Our ridiculous 55mph overall speed limit is not much impediment to anyone with a radar detector — but 100mph cruises on USA Interstates are hardly adventures to write odes about. Give me a country road and a nimble 327 Corvette anytime. Most of them — at least the non-injection variety — can survive on the low-octane fuels available, while

The 1966 Sting Ray convertible with 427 Mark IV V8 (note the special bonnet), capable of 0-100mph in 15 seconds. Despite the huge engine weight distribution was nearly perfect at 51/49 per cent.

Last of its generation, the 1967 Sting Ray coupe is arguably one of the best Corvettes ever built. All the styling cliches had been eliminated, and the range of power units was wider than ever before. Four-wheel disc brakes allowed it to stop as well as go; aluminium wheels were bolt-on type instead of knock-off.

The 1967 coupe, showing the egg-crate-style grille which had arrived in 1966.

The 1967 coupe with 427 V8 engine. The bonnet bulge for the 427 was different this year; the oblong reversing light was another new feature.

providing plenty of punch. The 300bhp version, which was offered in every year of the Sting Ray generation, is the best choice for a reasonable compromise between performance and economy. Naturally, this should be allied to the four-speed gearbox — as it usually is — and the leather upholstery option — as it usually isn't. Injected V8s are troublesome things, though one can't argue with the fact that they are highly prized by collectors. Big-block Sting Rays are highly specialized cars, suitable mainly for dragstrips and racing circuits — yet for those to whom cost is no object, they still provide a reminder of an age when horsepower was king, when we spelled 'performance' in cubic inches.

As to the choice between coupe and open Corvette, this is a highly subjective area. Bill Mitchell will say that the coupe — the original split-window 1963 variety, mind you — was the ultimate expression of the Sting Ray design. If you base your judgment on what the professionals say, the coupe is for you. Like the E-type coupe with which it was so often compared, the coupe is a ground-up styling job rather than a fastback evolution of the roadster, and therefore a highly unified and timeless-looking car. It is unlike anything else on the road.

The open Corvette always outsold the coupe, if only by a fraction in 1963, and it remains truest to the original concept of the car, and of sports cars in general. In contrast to numerous contemporaries from other countries, however, the roadster is anything but uncomfortable or draughty. The convertible top is a snug-fitting affair with proper weather-stripping, roll-up (or electric) windows and a fresh-air heater powerful enough for a limousine — sometimes coupled to factory air conditioning. The hardtop option merely adds an extra dimension (and about $1,000-$1,500 in value) to a Corvette roadster. Choosing between the open and closed version is basically easy, depending almost entirely on the way you like to take your motoring.

In terms of value, open Corvettes outpace coupes by about 10 per cent, all other conditions being equal — except for 1963, where the original 'split-window' model leads the roadster by about 20 per cent.

CHAPTER 6

Later Corvettes in competition

1963 to date

In the summer of 1962, Chevrolet Engineering began a project designed to head off Carroll Shelby's Cobras and maintain a Corvette dominance in sports car racing through the rest of the 'Sixties. They called the result 'Corvette Grand Sport'.

The original idea was to build 125 of these lightweight racing Corvettes in order to get them classified as Production cars, in the same sense that the Cobras were. In the end, they built only five during 1963 — but they were very impressive indeed. Grand Sports were mounted on a ladder frame with 6in frame rails; the suspension and transmission were based on conventional Sting Ray hardware, but were hand-fabricated in an effort to reduce weight. The four-speed gearbox was mated to Zora Duntov's experimental 377 cubic inch V8, which had an aluminium block, two spark plugs per cylinder and fuel injection. Horsepower was rated at 550 — more than enough, it seemed, to handle the Cobras.

The Grand Sport body was derived from the Sting Ray coupe, but in fact was smaller in all directions and equipped with certain functional air scoops and spoilers. Duntov's object was to race the Grand Sport team at Le Mans in 1963 — he anticipated achieving 180mph down the Mulsanne Straight — but luck wasn't with him. In March, Chevrolet decided that a factory Le Mans team would violate the spirit of the hoary old anti-racing resolution, and the Grand Sport programme was cancelled.

But what to do with the cars? Duntov decided to equip them with stock 327 V8s and 'sell' them to certain private parties long used to racing Corvettes. The privateers — Grady Davis, Dick Doane, Jim Hall and John Mecom — then 'hired' Duntov to help prepare the cars for competition. Later, Roger Penske and George Wintersteen acquired Grand Sport 'roadsters' — made by cutting off the tops of two of the coupes. Ostensibly a private venture, this was in effect a factory operation, since Duntov freely used the services of GM in fashioning the competition Grand Sports.

The intended 125 units not having been produced, Grand Sports could qualify only as Class C-Modified cars. This was a formidable racing class, dominated by the likes of Lister-Chevys and Chaparrals. But Grady Davis put Dr Dick Thompson behind the wheel of his Grand Sport, and Thompson finished a respectable fourth in the C-Modified category for 1963.

Also in 1963, Duntov managed to shovel his experimental 377 V8 back into a trio of Grand Sports for the Nassau, Bahamas Speed Weeks. Racing in the unlimited class against a bevy of Cobras, the Grand Sports triumphed, each of them gaining no less than 10 seconds per lap on the fastest Cobras. In 1964 the same three Grand Sports appeared at Sebring, but all dropped out with mechanical failure. Back at Nassau in 1964, Roger Penske won the unlimited class in his topless GS, while Wintersteen, Pete Goeltz and Ed Diehl teamed to finish 14th in a coupe. Wintersteen and Penske were still racing Grand Sports in 1966, though they were not truly competitive with state-of-the-art C-Modified cars by then.

The Grand Sport proved that unplumbed depths of performance lay within the Sting Ray package. Undoubtedly they would have proven more than a match for the Cobras had Chevrolet built the necessary number to qualify for the Production racing categories. Even without development work, Grand Sports were still showing their tails to formidable rivals in 1964 and were at least acceptable on the road courses as late as

Bob Johnson driving a shark-like Sting Ray at the Mid-Ohio SCCA Regionals in October 1966. The A-Production car was running a near-stock 427 V8. *(Dave Arnold)*

Jerry Thompson prepares to do battle with the dreaded Cobra . . .

. . . and Don Yenko proves it can be done, against Ed Lowther's 427 Cobra at Mid-Ohio Nationals in June 1966. *(Dave Arnold)*

1966. It really is too bad that GM and Chevrolet didn't give the Grand Sport the factory backing it deserved.

Without the GS, SCCA's Class A-Production continued to be a Cobra parade, although John Martin, from Florissant, Missouri, did interrupt the procession by winning the Midwest Division A-Production championship in 1965. In 1966, Duntov introduced the L88 engine (427 cid, 560bhp in competition tune), which combined with the now-famous F-41 suspension, heavy-duty brakes and Positraction rear axle to produce what looked like a formidable package. But the Cobras had almost as much horsepower and were lighter by half a ton. It just wasn't enough.

Not that GM wasn't giving any support. At least covertly, they were heavily subsidizing and assisting Can-Am racers like Jim Hall and the Camaro Trans-Am team of Roger Penske. They had basically made a decision to abandon SCCA classes to the Cobras because the Can-Am and Trans-Am had more potential — both for winning and for good publicity. Given the restrictions of the General Motors bureaucracy, it was impossible for Chevrolet Division to support an outside subcontractor in the way that Ford was supporting Shelby. It was equally impossible to turn a mass-produced sports car like the Corvette into a competition champion.

In the long-haul enduros, however, it was another story. In 1966, for example, Penske, Wintersteen, Dick Guldstrand and Ben Moore combined to finish 12th overall and first in the GT class with a Corvette at the Daytona Continental. The same team ran first in class and ninth overall at Sebring; Dave Morgan and Don Yenko repeated the performance at Sebring in 1967. And at Le Mans, Bob Bondurant and Dick Guldstrand led the highly competitive GT class for many hours. They looked like sure winners, but engine trouble retired them.

Production of Shelby's Cobras was wound up in 1968 as the first Federal emission and safety regulations took hold. In the same year, Chevrolet introduced an all-new Sting Ray with dramatically better aerodynamics and a cold-air induction option for the L88 V8 engine (see Chapter 7). The roadster version with lift-off hardtop turned out to be more slippery than the coupe; Corvette was on the verge of another championship era.

The first new models appeared in competition at the Daytona Continental in February 1968, when Dave Morgan and Jerry

Trundling goliaths. A pair of A-Production 'Vettes, including a split-window 1963 Sting Ray, oversteer and understeer through the esses at Candlestick Park in June 1965. *(Corvette News)*

Grant placed 10th. At Sebring, three new Corvettes were entered in the Group 3 GT class, which was won by Morgan and Hap Sharp. It was the first time in years that Corvettes had soundly trounced the Ford-powered cars, though we have to admit that there were few Ford-powered cars to contend with. But the Corvette was improving, too. An aluminium-block version of the L88, called the ZL-1, was made available in 1969, and Corvette's first SCCA Production title since 1962 was at hand.

Instead of a factory-backed team, Corvette drivers now began to rely on sponsors from outside the automotive area. One of these was Owens-Corning Fiberglas, builders of the raw material for the Corvette body. In 1969, Owens-Corning's Jerry Thompson was an easy winner in A-Production, while Allan Barker nailed down B-Production in a 327 Corvette. It looked like a bright new era for SCCA Corvettes, but now a new problem was at hand — the age of Federal control over automotive design.

As Zora Duntov neared retirement and Federal regulations became stiffer, the impetus to design and build 'works' racing Corvettes slackened appreciably. But you could still buy a racing option, equivalent to the old RPO 684 competition suspension, and a new 454 V8 called the LS-5 was available to go with it. Suspension and engine were combined to form the ZR-2 option. There was also a ZR-1 option, using the solid-lifter, high-performance 350 V8 (LT-1) engine and the competition suspension.

The number of independent sponsors burgeoned in the 'Seventies as officially Detroit moved away from direct involvement with racing. In addition to Owens-Corning's cars, actor James Garner sponsored a team of American International Racer Corvettes that did reasonably well. But the real star of the 'Seventies for Corvette was John Greenwood . . . driver, builder, engineer and promoter.

Greenwood was something of a prodigy. In 1968, he attended his first SCCA driving school; in 1970, he won the A-Production national championship. In 1971, he won it again, and also teamed with Dick Smothers to win the GT class at Sebring. Greenwood's meteoric success made him an attractive property; in 1972, he raced on Goodrich T/A radial tyres and was supported by the Goodrich company. At Le Mans in 1972, Greenwood's Corvette qualified at a higher speed than any other GT class car, and was

The awesome 1963 Grand Sport, only five of which were built as a possible challenge to the Cobras. Original engine was an experimental aluminium-block 377 V8 with fuel injection and a claimed 550bhp gross. Duntov predicted 180mph on the Mulsanne straight at Le Mans, but Chevrolet cancelled the project as an infringement on the AMA anti-racing 'ban'. *(Dave Gooley)*

The purposeful instrument layout of the 1963 Grand Sport. An engine oil temperature gauge is the top dial on the centre console, with fuel pressure and differential oil temperature gauges below, separated by switches for the primary and secondary fuel pumps. *(Dave Gooley)*

Four dual-choke carburettors nestling on top of the aluminium-block 377 V8 engine of the Grand Sport. Note the ample space between the auxiliary drive pulleys and the radiator. *(Dave Gooley)*

J. Marshall Robbins' 427 Corvette in the Trans-Am race at Road Atlanta in April 1973. *(Dave Arnold)*

The Owens Corning Team's Tony DeLorenzo drove this 427 Corvette at Daytona in 1971. *(Dave Arnold)*

well ahead of all the rest when his LT-1 tore itself to bits. In 1973, Greenwood and Ron Grable finished third overall at Sebring in one of Corvette's finest performances.

John Greenwood helped reorganize Sebring in 1975, after the energy crisis had threatened to put an end to it forever. But the new Sebring was a place for Porsche GTs, and though Corvettes continued to place well there, all the winners were built in Stuttgart. Similar changes occurred in Trans-Am and IMSA racing, where the Corvette increasingly became large and unwieldy to compete with the Porsches. Greenwood spent $100,000 on each of his 'Profile' Corvettes for Formula 5000 — they used spaceframes, hand-fabricated suspensions and swoopy, aerodynamic bodies — but even with 220mph on tap they weren't fast enough to stay with the twin-turbocharged Type 935 Porsches.

In SCCA road racing, however, Corvette maintained its dominance after the departure of the Cobras. In A-Production, Corvettes triumphed every year save 1973, when Sam Feinstein succeeded with an aging Cobra. In B-Production, they won every year except 1975, when Bob Tullius drove an E-type Jaguar V12. The most remarkable B-Production car was the Stingray raced by Allan Barker, which won the championship four years running in 1969-72, was sold to Bill Jobe, and then won the 1973 and 1974 championships as well. The change of character which accompanies the new generation cars of 1984 to date has not been particularly angled toward competition, but SCCA racing goes on and ought to.

Throughout its history, the Corvette has sustained a very impressive competition record. There was, admittedly, always something around that was a little bit quicker. The Mercedes-Benz 300SLs were too much for the Corvette in the mid-'Fifties; the Cobras generally outshone the Corvettes in the 'Sixties, if only because GM was less interested in road racing than Ford; and the Porsches have proved hard to catch in the 'Seventies and 'Eighties. Yet through it all Corvettes have continued to race consistently with the leaders. Counting all the seasons and classes, good years and bad, the sheer volume of the Corvette record compares with those of Lotus and even Ferrari — and one could argue that the Corvette has been a lot more flexible as a competition car than either of those. Beyond the road courses, Corvettes by the thousands have been used with success in drag racing and solo events; the ubiquitous Chevrolet small-block V8 has itself powered everything from competition power boats to Indianapolis cars. Thus every driver of a Corvette today may take pride in the knowledge that what's under him is a thoroughbred — a car proven on the toughest courses, against the most formidable competition around.

CHAPTER 7

Stingray and a changing image
1968 to 1983

Chevrolet decided to issue no 1983 model year Corvette, but to bring out the all-new 1984 model early. In some ways you could call this decision a valedictory — tribute to the longest-running and most successful generation of Corvettes in history. If the new model does as well, GM will be more than satisfied.

As Mitchell said, plans for a fully revised Corvette were laid down almost immediately after the 1963 Sting Ray made its appearance. Concepts followed two general themes. One was a mid-engine car, with sharply raked front end, broad expanses of curved glass, skirted rear wheels and a periscope replacing the rear window. The high cost of the mid-engine transaxle, rather than the extensive restyling this car would have required, ultimately killed this project. The second and successful design route was a conventional front-engine car based on the experimental Mako Shark II.

The designer most responsible for the 1968 production Corvette was David Holls, a car enthusiast as well as a highly respected GM stylist. Holls took pains to come up with something acceptable not only to the higher brass, but which would be a suitable replacement for the handsome 1963-67 generation. He retained the flat, ground-hugging snoot of the Mako Shark II, but instead of a fastback roofline, the production roof was notched and the rear deck ended in a Kamm-type panel with a built-in spoiler.

To the surprise of just about everyone, this shape was not nearly as slippery as it looked. Wind-tunnel tests suggested that the 1963-67 Corvette was actually more aerodynamic than this proposed replacement. The new car was to have appeared for 1967, but Zora Duntov asked Pete Estes to hold off for another year while the wind-tunnel tests continued. Accordingly, the front fenders were cut down, the notchback was redesigned, and a less prominent rear spoiler was adapted. Rear glass was broadened to improve visibility and front wing louvres were designed to assist cooling as well as to decrease wind resistance. The slightly raised front end received an air dam and important detail features were pop-up hidden headlights under individual bonnet panels, and hidden windshield wipers.

The 1968 production models used the same engines as the 1967 cars, including the 'ultimate' L88, which had been announced in mid-1967. This all-out racing engine produced up to 560bhp at 6,400rpm, gulping 103-octane fuel to satisfy its 12.5:1 compression ratio. The L88 featured aluminium heads, larger valves, an aluminium intake manifold, a Holley 850 carburettor and a small-diameter flywheel with heavy-duty clutch. In 1969 it was joined by the ZL-1 racing engine with dry sump and aluminium block. The ZL-1 weighed 100lb less than the L88, but it was a $3,000 option. (Because these were not truly 'street' engines, they are not shown in the engine charts accompanying this chapter.)

This new Corvette has only recently been considered by collectors. Their interest has as much to do with the passage of time and rising prices of pre-1968s as it does with the car's intrinsic value. Many Corvette collectors hold that the 'classic' Corvettes stopped after 1967. Mixed opinions about the new car were recorded from the day of its introduction, too. In the opinion of *Road & Track*, it was 'highly reminiscent of certain older Ferraris, laid around a chassis that seemed fairly modern in 1962 but is now quite dated by the march of progress . . . the

Designed largely by David Holls, the 1968 Corvette was based on the Mako Shark II showcar (see next chapter). For this year only, the model designation was 'Sting Ray' instead of 'Stingray'.

The roadster or convertible version of the Sting Ray for 1968. With this generation, the fully open body style began to recede in buyer favour from the coupe, which now had removable roof sections. The last fully open Corvettes were built for the 1975 model year.

The 1968 coupe with roof panels in place and removed. Holls' styling, ultra-clean at first, gradually became cluttered as the years went by. In its initial form, like many designs, it was fresh and aggressive.

general direction of the change is away from Sports Car and toward image and Gadget Car'. This view was widely recorded by the 1968 motoring press and you still hear it today in Corvette circles. The problem was that the 1963-67 generation had been so good that its replacement represented something of an anti-climax. An analagous situation is the succession of the E-type Jaguars by the XJ-S.

Critics had plenty of solid facts to back up their contention that the 1968 was a disappointment. The body, they pointed out, was 7in longer than the 1967, most of those inches occurring in front overhang. The wheelbase was unchanged at 98in, but the interior was more cramped and there was less luggage space. About 150lb had been added to what *Road & Track* called the 'already gross avoirdupois'. The '68 Corvette was in the editors' opinion perfect 'for those who like their cars big, flashy and full of blinking lights and trap doors . . . The connoisseur who values finesse, efficiency and the latest chassis design will have to look, unfortunately, to Europe'.

General Motors has a remarkably good ear for so large a corporation, and opinions such as those of *Road & Track* were duly noted. By 1969, the Corvette had been returned to 'separate but equal' status within Chevrolet Division and Zora Arkus-Duntov, temporarily shunted aside, was restored to a position of authority over future model development. Whether it was Duntov's doing or not, the 1969 version again bore the one-word designation 'Stingray', which had been rendered as 'Sting Ray'

Close-up of the fast tapering sail panels and the deeply recessed rear window of the 1968 Corvette Sting Ray.

for 1968.

There wasn't anything they could do with the basic package, of course, but some obvious and needed janitorial work had been done with the details. Exterior door handles were cleaned up, black-painted grille bars replaced chrome and the back-up lights merged with the inner tail-lights and looked much simpler. The interior had more passenger space and cargo stowage area. Handling was better by virtue of wider-rim wheels and a considerable amount of vibration was eliminated through a stiffer frame. The smaller engine was stroked to 350 cubic inches and offered with 300 or 350bhp. Four 427 cid engines were offered along with an assortment of axle ratios, which ranged from 4.56 to 2.75:1.

Though these good efforts weren't lost on enthusiasts and the press, there was still plenty to complain about. The Corvette, *Road & Track* now ventured, 'lacks finesse; it's like using a five-lb axe when a rapier, properly designed, could do as well . . . The person we associate with the Stingray is the Animal, one who prefers to attain the goal with brute strength rather than art and fast footwork.' But *Car Life,* then published by the same company, was kinder. While admitting these faults, plus a tendency 'for things to fall off', *Car Life* insisted that 'Corvettes are for driving, by drivers . . . The Corvette driver will be tired of smiling long before he's tired of the car'.

Of course the main purpose of any GM product is to make money, and this the new Corvette did. An all-time high had been set in 1968 with 28,566 units for the model year — it lasted only until 1969, when the figure reached 38,762. The 1970 models were largely unchanged except in price — the bottom line now exceeded $5,000. Sales dropped substantially, only 17,316 of the 1970 models being built, with coupes leading convertibles by a

Good use was made of the space behind the seats in the provision of a glove box and a larger stowage well, the lids of which were finished in carpet material.

Chief identifying distinction on the 1969 Stingray was the blacked-out grille, which gave it a more functional appearance up front. Out back, the reversing lamps merged into the tail-lamps. Marker lamps at the extreme front on either side used amber lenses for the first time — the 1968 markers were clear.

five-to-three ratio. But in 1971 and 1972 sales picked up again to 21,801 and 26,994, respectively.

The new 454 cubic inch engine reflected increasingly tight Federal emission controls, and was the replacement for the 427, which saw its last use in the '70 Corvettes. With a bore and stroke of 4.25 x 4.00in, it was designed to emit fewer pollutants and to run on lower-octane petrol. A more powerful engine with 465 horsepower was to have been offered, but it was withdrawn because it could not be made to meet the maximum emission level. The solid-lifter, small-block LT-1 engine, producing 370bhp on 11:1 compression, was likewise eliminated after 1970 for emission reasons. The effect of Federal regulations on regular Corvette engine options is dramatically shown by the following summary:

Small-block V8s (350 cid except 1968's 327)

1968
350bhp @ 5,800rpm, 11:1 CR

1969
350bhp @ 5,600rpm, 11:1 CR
300bhp @ 5,800rpm, 10.25:1 CR

Not much to fault here. Easily readable instruments with the jumbo speedo and rev-counter well buried to eliminate glare, and minor gauges well positioned in the console. No glove box, though, on this 1969 model.

1970
370bhp @ 6,000rpm, 11:1 CR
350bhp @ 5,600rpm, 11:1 CR
300bhp @ 5,800rpm, 10.25:1 CR

1971-72
330bhp @ 5,600rpm, 9:1 CR
270bhp @ 4,800rpm, 8.5:1 CR

Mark IV V8s (454 cid 1971-on; others 427)

1968
435bhp @ 5,800rpm, 11:1 CR
400bhp @ 5,400rpm, 10.25:1 CR

1969
435bhp @ 5,800rpm, 11.1 CR
430bhp @ 5,200rpm, 12.5:1 CR
390bhp @ 4,800rpm, 10.25:1 CR

1970
390bhp @ 4,800rpm, 10.25:1 CR

1971
425bhp @ 5,600rpm, 9:1 CR
365bhp @ 4,800rpm, 8.5.1 CR

1972
365bhp @ 4,800rpm, 8.5:1 CR

Nineteen seventy-two marked a definite turning away from horsepower and performance, and towards low emissions achieved by detuning — a process that would continue throughout most of the 'Seventies and make a whole generation of

A 1970 Corvette with the 427 engine, seeing its last use in this model year. Styling became busy again with egg-crate grillework. Sales were off and only about 17,000 '70s were sold.

cars far less interesting than their predecessors. Only in the 'Eighties, as the American industry has begun to develop turbocharged small-displacement engines of fewer than eight cylinders, has 'performance' become an acceptable word again in Detroit.

On the 1973 Corvettes there were no mechanical-lifter engines, and no LT-1s. Horsepower ratings were in net bhp (previous figures had been gross), and if the numbers were more accurate, they were a lot less impressive. The 350 V8s rated at 270 and 330bhp in 1972 were now listed at 190 and 250 net horsepower. The sole remaining Mark IV engine was at 9:1 compression with 275 net horsepower.

Physical changes in the Stingray through the 1970s were relatively small. In 1972, a burglar alarm became standard — ominous testimony to the rising rate of Corvette thefts. (An insurance executive told the author that in actuarial terms, the buyer of a new Corvette in an urban area has a less-than-even chance of keeping it for the first year!) For 1973, removable roof

Styling revisions for 1970 included 300SL-type front wing louvres, a more complicated design than the original. Square exhaust outlets replaced round items, having proven more efficient and of higher capacity.

Stingray seats were thin and not too buckety. Leather was a popular upholstery option; when ordered, it came on seat surfaces only. There was an awful lot of moulded plastic in the cockpit of this '70 model.

panels were temporarily eliminated from the coupe because Chevrolet anticipated roll-over standards from Washington. These didn't materialize and the roof panels went back on to the option list in 1974. A mandated change (in force for 1974) was anticipated by the application of a 5mph bumper to the 1973 model. It was made of polyurethane and painted the body colour.

The temporary comeback of Duntov and those who felt as he did about cars was now history, and in those meagre years for enthusiasts the Corvette was a pretty dull package. It was designed to appeal to a different market from that which had traditionally bought the cars — and it was considerably more expensive. The 350 V8 with 250 net bhp had a base price of $5,500 in 1974, and usually sold for about $750 more. Kerb weight was up to 3,500lb, and comfortable touring was stressed instead of all-out performance. Yet this milque-toast product — to the annoyance of people like motoring writers — sold well. A high of 49,213 Corvettes was hit for the 1977 model year, though the Arab oil embargo of 1977-78 put a cap on sales and it would take the marque a long time to approach that figure again.

The 1974 models continued with minor, evolutionary changes, mainly to meet the government regulations. The Kamm-back tail end was replaced with an energy-absorbent, rounded shape that

The convertible Stingray for 1970. In a low-production year, this model saw only 6,648 copies, and therefore ranks as one of the more desirable open Corvettes of the 1968-75 period. Look for it!

was less effective at holding the rear end down at high speeds — this was considered little loss since high speeds seemed to have gone out of America like the proverbial one-hoss shay. Chevrolet also saw no need for the 140mph-rated tyres, and changed down to 120mph ratings the same year. They did offer an optional 'gymkhana suspension', with the ancient handling-improvers of high-rate springs and firm shockers — though it's hard to imagine a 1974 Corvette giving much trouble to anything else on a gymkhana course. Luxury options sold much better — electric window lifts, vacuum-assisted brakes, integrated air conditioning, stereo tape decks and leather upholstery were all popular.

Things would get worse before they'd get better. In 1975 the Mark IV V8 disappeared from the option lists, leaving a sole 350 which produced varying anaemic horsepower through to 1980. Here's the way it went:

Year	bhp (opt)	CR (std)
1975	165 (205)	9.0:1

The Stingray coupe for 1971, another low-volume year, and the first for the new emission-controlled 454 Mark IV V8, which offered up to 425bhp for 1971 only. Few styling changes occurred.

1976	180 (210)	9.0:1
1977	180 (210)	9.0:1
1978	185 (220)	8.9:1

All this is relative, of course, because compared to the rank-and-file Detroiter of the late-'Seventies the Corvette was still a pretty impressive car. And if you can sense a certain upward trend in the power ratings above, you have realized that the United States Government aside, Chevrolet was still in there trying. In 1978, the Corvette's main change was an extra modicum of rear glass, a new glovebox and a larger tailpipe. Yet the 350 V8 had more vigour than in 1977, with the standard L-48 and optional L-82 versions receiving 5bhp and 10bhp boosts, respectively. The L-82 (not sold in California, where emission control was tighter than in the other 49 states) breathed more easily with its dual-snorkel air intake, larger exhaust tubing and freer-flowing mufflers behind its catalytic converter. It delivered 0-60mph in 6.5 seconds and reached the quarter-mile in 15 seconds at 90mph, which was hardly tarrying.

Two 1978 models which may prove to be worth collecting (the verdict won't be in for another five years at least) were the Silver Anniversary and Indy Pace Car Corvettes — in Bill Mitchell jargon, 'tape stripe packages' — designed to help boost sales in the first year of the current recession (which, as recessions go, is

The '72s were again little changed, but this was the last of the chrome-bumpered models, and in many ways it marked a turning point in Corvette performance. By 1973 there wasn't a hot engine left.

getting to be a grand-daddy!). Mitchell likes silver paint on anything from cars to bread boxes, and since the Corvette was celebrating its 25th anniversary this seemed natural. The Silver Anniversary option included 'anniversary' badges, which decorated a silver upper body and grey underbody, highlighted by a pinstripe along the dividing line. Except for the paint and badges, however, the Silver Anniversary Corvette was depressingly 'stock' for a celebration car. Whether these decorations ever render it an above-average proposition for the Corvette collector is a question still in doubt.

Another PR gimmick resulted in the Corvette being 'chosen' to pace the 1978 Indianapolis 500. Chevrolet Division duly took advantage of this to build 2,500 Pace Car replicas for sale to the public on a first-come, first-served basis. The only problem was that there are 6,200 Chevrolet dealers in the United States — so in the end, 6,200 'Pace Cars' were built. The option was RPO Z78.

Like the Silver Anniversary Corvette, the Pace Car used a two-tone paint treatment: black on the upper body, silver metallic below. The dividing pinstripe was bright red. Alloy wheels and jumbo radial tyres were standard; the removable roof panels were glass. The Pace Car interior was pure Bill Mitchell — silver leather or leather-and-cloth with matching silvery grey carpets. A

Pre-production '73 Corvettes on the GM test track. The cars were new-looking, with 5mph bumpers of body-coloured polyurethane, and removable roof panels were dropped for this year only. Handling was pretty much the same, though; good cling, but plenty of skill always required when powering through the bends.

Pretty decorations for a slower Corvette. A polyurethane-covered bumper transformed the front end; Chevy had also placed more emphasis on sound insulation and a smoother ride. This was the last year for 140mph-rated GR70 × 15 radials.

Smelly plastic cockpit of the 1974 Corvette. Things had always been pretty cramped in there, and manhandling luggage past the seats was a traditional problem. Removable roof panels were back.

The 1974 Stingray coupe could be ordered with 'gymkhana suspension', which gave tenacious sticking power amidst considerable roll plus creaks, groans and other bumps-in-the-night. This was the last year for the Mark IV big-block V8, with which this car seems to be equipped.

Pretty much the same formula was used in 1975, the last year for the fully open convertible. Small bumpers were added at front and rear. Horsepower hit a low spot for Corvette V8s; only 165 net was standard. To many, the '75 was underwhelming; all show and no go.

Little change again for 1976, when the sport coupe became Corvette's sole body style offering. Light-alloy wheels were both attractive and functional. Horsepower now 'soared' to 180, with 210 optional.

dramatic feature was a hefty front spoiler similar to that of the Firebird Trans Am, which wrapped around and blended into the wheel wells. The rear spoiler looked 'tacked-on', in contrast, wrapping only part of the way down the body sides. Pace Car replicas also came with power windows, rear demister, air conditioning and sport mirrors, as well as tilt steering wheel, heavy-duty battery, power door locks and an AM-FM stereo radio with either 8-track tape player or CB. The cars were delivered with a packet of tacky decals, which the owner could mount on the doors. They read: 'Official Pace Car, 62nd Annual Indianapolis 500 Mile Race, May 28, 1978'. Compared with the $9,531 base price stock Corvette, the Indy Pace Car cost $13,653. This was not enough to quelch demand, however, and a lot of dealers marked them up, some asking over $20,000. One arrant optimist advertised a Pace Car for $75,000 in the US vintage car press, no doubt in an attempt to test the 'greater fool' theory.

As 1978 passed into history, General Motors still wasn't sure what to do with the Corvette in future. Dozens of proposals and showcars had been touted over the years as the long-awaited replacement to David Holls' 1968 design (which was actually begun in 1963 and whose chassis was clearly related to the 1963 production car). A flurry of hoopla surrounded the exciting Aerovette, a splendid one-off, and among the last of Bill Mitchell's specials before his retirement. But the Aerovette was ultimately shelved.

The decision, as we now know, was to keep on selling the old design as long as it would sell. This turned out to be through 1982 — an astonishing record for any automobile, and particularly a sports car.

The 1979 Corvette was, accordingly, much the same as the 1978 model, which had introduced the new 'glassback' rear window treatment. However, the '79 did feature black-finished mouldings on the roof panels and rear window, and the old cross-flag medallions were reinstalled for the sake of tradition.

As for the engine, the 350 cid V8 (now beginning to be referred to as the 5.7-litre) was carried over with minor improvements, which increased torque and horsepower, along with cold-engine drivability. Manual-gearbox models now used the same shock absorbers as the automatic, giving them a more comfortable ride at some sacrifice in handling. The rear axle ratio was increased

Another design period is ending here, with the 1977 Stingray. It was the last of the original Holls-designed notchback bodies, destined to reappear for 1978 as a 'glassback'. The power rating was unchanged.

numerically from 3.08 to 3.55:1. High-back bucket seats, introduced on the limited edition 1978 models, were now standard. The car still lacked any exterior access to the luggage compartment, so Chevrolet hinged the backrests to allow the seat backs to be folded flat, level with the luggage area for loading.

The last year for the old 350 V8 was 1980 — a time of 'major surgery' on the old body. Chevrolet finally responded to the need for better fuel economy by trimming over 250lb from the kerb weight. This dropped the Corvette two weight classes (by Environmental Protection Agency standards), aiding Chevrolet's effort to meet a corporate average fuel economy (CAFE) standard of 20mpg. With the weight reduction, the one-piece bumper-grille had an integral air dam, said to improve aerodynamics and engine cooling; a more deeply recessed grille; and a new glass-fibre face bar and corner braces, instead of the metal ones used on the 1979 front bumper. The rear spoiler was integral and the tail-lamps were redesigned. Chassis changes included revised frame and 'birdcage' members and a new transmission crossmember, both lighter than in 1979. The differential housing and cross

A made-to-order collectible, or so the promoters claimed, the Indy Pace Car replica was an outlandishly painted Corvette with a set of decals for your doors (putting them on was left to the buyer). Approximately 6,200 were built, all loaded with goodies like power windows, air conditioning, AM-FM stereo with tape deck or CB radio, and the special black-and-silver paint job. Base price was $13,653, and demand was such that dealers sold many for more than this.

supports were now made of aluminium, and the exhaust system was redesigned to save weight and improve performance. Powertrain revisions included new gear ratios and a torque converter clutch for all automatic Corvettes. The latter engaged automatically at 30mph, or when the brake pedal was depressed. The four-speed gearbox received numerically higher ratios on first and second, allowing the use of a numerically lower rear axle ratio with no performance penalty.

The 1981 Corvette was again a mildly revised successor model, benefiting from the calorie-counting programme of 1980. But the long-running L-48 and L-82 V8 engines were gone, replaced by the new L-81 engine of the same capacity. The L-81 featured GM's Computer Command Control electronic emissions control system. CCC also controlled engagement of the automatic gearbox's lock-up torque converter, effective on second and third gears. A few more pounds were shaved via a GRP-reinforced monoleaf rear spring, which replaced the previous steel multi-leaf spring dating back to 1963. Thinner door glass and a lighter stainless-steel exhaust manifold were also new in 1981, while six-way power seats were added to the option list. Standard was a starter-interrupt feature, added to the alarm system, which prevented engine starting after a forced entry, even if the ignition was bypassed.

The 1982 Corvette is probably destined to become a collector car a few years down the line. Not only does it represent the end of an era, it also ushers in the new. Under the skin it features an entirely new drivetrain — the one to be used in the all-new 1984 model.

Traditionalists will be reassured in that the '84 — like the '82 — has retained a V8 powerplant and, while the Camaro has gone to the 305 cid (5-litre) V8, the Corvette still has the 350 (5.7-litre). The induction system is all new — instead of carburettors, Chevrolet now mounts Throttle Body Injectors (TBIs) atop the dual-carb intake manifold. These represent a sort of space-age fuel injection, metering fuel into the airflow electronically. Chevrolet calls this feature 'Cross Fire Injection', which is a rather unfortunate name. (One *Road & Track* tester said it sounded more like a malfunction than a sales feature: 'Stand back, kid, that engine's about to cross-fire!'.) But there's no doubt that TBI is an important improvement on the carburettor.

To celebrate the Corvette's Silver Anniversary, Chevrolet also produced this commemorative model for 1978. It featured a special paint job, approximately opposite the Indy Pace Car's, and a few custom badges. Marking an important historical milestone, it will be a collectible car in future years.

Whereas 'glassback' styling and special models made the news for 1978, 1979 was a year of minor refinements. High-back bucket seats introduced on the Pace Car were made standard and they pivoted higher to allow easier access to the luggage compartment.

The Stingray for 1979, which boosted optional bhp to 225 net on the L-82 V8, had more suds than the marque had seen for five years.

Compression is back up on the 1982-84 V8 — to 9.0 instead of 8.2:1. The camshaft has been redesigned with better performance in mind, featuring higher lift and overlap. Stainless-steel exhaust headers are new, and emissions are handled via an exhaust-gas recirculating system, evaporative control and a huge catalytic converter. The result is 200 net bhp at 4,200rpm and 285lb/ft of torque at a low 2,800rpm — 'A far cry from the 400bhp-plus days of the L88 and L-68', said one road-tester, 'but not exactly a shrinking violet by today's wheezing standards'. A late-1982 test of a Corvette with automatic showed 0-60mph acceleration in 7.9 seconds and the standing quarter-mile in 16 seconds at 85mph. This is a definite improvement from 1981. Combine that with the fact that the new '84 model weighs 500lb less and we can see the progress that is being made.

Though the 1982 Corvette was offered only with four-speed automatic transmission, the manual option has returned for 1984. The automatic is governed by a black box — the Electronic Control Module — which also decides how to govern the TBIs, water temperature, throttle position, oxygen level and engine knock. ECM 'decides when the torque converter should lock up, providing a standard-gearbox-like mechanical drive when load conditions are right'. Numerous bugs have been reported on the 1982s, with transmissions failing to shift when they should and locking up at less than opportune moments, so we should hope that all these nits have been combed out of the 1984 model.

Marking the end of a long run, the 1982 Corvette was offered in a Collector Edition, a special which accomplished too late what was needed for years — exterior luggage access through a lift-up rear window. Production has been shifted, now, from the old St Louis factory to a new, modern plant in Bowling Green, Kentucky. This has apparently contributed to improved fit and finish, though the '82 still rattles and shakes with all the aplomb of the '68. Lack of torsional rigidity has been endemic to the breed since then, and another thing we can hope is that they'll get

Wind-tunnel work dictated changes to the Stingray for 1980, and the engineers also managed to shave off 250 pounds. A new option was a roof panel carrier, which attached directly to the body.

the shakes out of the animal by 1984.

Driving the 1982 Corvette is a familiar experience. The car sticks well on corners, but is anything but nimble, and oversteer can be readily induced at the limits of adhesion. The driver is buried bathtub-style in the cockpit, and the space for both occupants is still pretty limited. The ride is hard, too much unsprung weight still being carried on a very tight suspension.

There's plenty of noise — engine, exhaust, wind, body creaks — and fuel mileage is the usually dismal 13mpg (US gallons), stretchable to perhaps 18mpg on a trip.

The 1984 Corvette (if any 1983s are sold they will be 'retitled' 1982s) represents an important improvement in virtually all respects. Thus it will be as significant in Corvette history as the 1956, and the 1963. Perhaps it is the most significant Corvette of

Cleaner than it had looked in years, the '80 Corvette was quicker and more slippery than before, but the manual gearbox had wider ratios in the interest of better fuel economy.

Aerodynamics were unchanged for 1981, but still more weight had been pared, and the new L-81 engine arrived, with its Computer Command Control emissions system. A neat new feature: you can't jump-start an '81 Corvette after breaking in, thanks to a standard-equipment starter-interrupt system.

Last of a long line, the 1982 Stingray; one of the best finished Corvettes ever, thanks to a modern new Kentucky assembly plant. The 5.7-litre V8 was hooked to a four-speed overdrive automatic transmission as standard. The four-speed box has returned in 1984 models.

You'd think they'd have learned! Collectors have traditionally shunned ready-made 'Collector Editions', causing values to plummet after the initial hoopla and sheep-shearing efforts on the part of dealers. We certainly don't recommend you buy any '82 as a potential collector's item — it's too soon to judge them. An important feature of the 1982 Corvette CE is the frameless glass hatch, which lifts for access for the first time. They should have done it years ago.

The sharp end of the 1982 Stingray, the Cross-Fire throttle-body fuel injection of which was to be transferred intact to the new 1984 Corvette and surrounded by a new body and chassis. Some '82s may be retitled '83s to stave off customer complaints, but officially 1983 has been skipped.

all time, its style and performance better than ever. And despite the imperatives of corporate bean counters, the mandates of Congress and the machinations of the Arabs, that's what Chevrolet has been trying to do all along. In 1984, for the first time perhaps since the mid-'Sixties, they're being given a real opportunity.

It hasn't been easy, for sure. In a 1977 interview with John Lamm, Zora Arkus-Duntov summed up his 25 years' experience with the marque in one word — 'struggle'. When John asked him, 'With whom?', Duntov replied with a one-liner that said it all: 'With whomever came to have an opinion different than mine'.

CHAPTER 8

Showcars and specials

The might-have-been Corvettes

The origins of the Corvette are closely intertwined with the 'dream car' programme of General Motors Styling and its postwar extravaganzas of glitter, girls and gladiolas, the GM Motoramas. These incredible circuses, which began in 1949, actually did serve a purpose other than publicity. GM 'plain clothesmen' freely circulated through the crowd, listening to (and writing down) people's reactions to the latest showcars from the den of Harley Earl. Earl himself often walked the floors, and his system must have been good, because in product design he rarely gave the public something they didn't want.

The first Corvette was a Motorama showcar, and though the Motoramas ceased in the early-'Sixties (when GM began suspecting its rivals of stealing its design ideas), the same process goes on today. Every new generation of Corvettes, including the '84s, has been preceded by at least one showcar, the *chief* function of which has been to test public opinion.

At the height of his powers in the early-'Fifties, Harley Earl was personally able to choose what GM would feature at each year's Motorama. For the 1953 show, as we already know, Earl's number-one special was the prototype Corvette. The initial work had been assigned to stylist Bob McLean, who began by drawing the rear wheels. Then he drew the seats, as close to the rear wheels as possible; finally he plugged in the firewall and engine, placing the latter as far back and as low as possible for optimum weight distribution. McLean's drawings determined the wheel base, which turned out to be 102 inches.

Most Motorama showcars were running automobiles, not simply custom bodies built over wooden bucks, like those of many contemporaries. The showcars used conventional passenger car chassis, which explains why some of those swoopy two-seaters like the Buick Le Sabre look a bit out of proportion; underneath was a full-size saloon chassis-frame. For the Motorama Corvette, however, Harley Earl ordered at considerable expense a special-wheelbase chassis. That this expense was approved is a testimony both to Earl's high standing and GM's intention to make the Corvette a production car.

The Motorama showcar which thus made its debut at the Waldorf Astoria, in New York, in January 1953, had somewhat different trim from the later production cars, and special wheel covers. But its Polo White paint job and red upholstery was an exact match of the 1953-54 Corvettes. Public reaction to the Motorama special was overwhelmingly enthusiastic and this was duly reported back to GM President Harlow Curtice by Harley Earl through his team of 'Motorama moles'; the Chevrolet sports car was quickly scheduled for production.

For the 1954 Motorama, Earl's team confidently prepared no fewer than three showcars based on the now-production Corvette. The first was a modest improvement on the standard roadster, featuring genuine wind-up glass side windows and a bolt-on hardtop. These improvements were so obviously useful that they had no difficulty entering into the scheme of things for 1956.

The second 1954 special was a radical fastback coupe on the stock Corvette body, labelled the 'Corvair' (as recorded earlier, this was a combination of 'Corvette' and Chevy's top-line model, the 'Bel Air'). The Corvair featured Earl's 'shadow-box' number-plate nacelle at the rear in the form of a broad arch trimmed with bright metal. The bodywork from the cowl forward was pretty much stock-Corvette. Earl and Ed Cole wanted Chevrolet to add

Responsible for every showcar from the original Stingray to the Aerovette, now-retired William L. Mitchell has left a great legacy. Here is Bill with the Mako Shark II and his personal Stingray.

The original Corvette as shown at the 1953 GM Motorama. Compare this car with production models shown in Chapter 1. The most notable change is to the side trim, but there are also: chrome (not plastic) inside release knobs, chrome shift knob, two extra dashboard knobs on passenger's side, no 'flippers' in forward upper deck mouldings, no windshield-end seals, no drip mouldings, upper front wing scoops, narrow headlamp bezels, front name-plate, oversize wheel spinners mounted 90 degrees from stock, rear deck name-plate and no upper dashboard edge vinyl.

this handsome coupe to the Corvette line as a regular production model, but the slow sales of the contemporary open cars caused this programme to be shelved.

The third Motorama car for 1954 was the most significant, though the only one of the three not built on a 102in Corvette chassis. This was the Corvette Nomad, a sporty estate wagon designed by Carl Renner of the Chevy Studio. It rode the 115in wheelbase of the standard Chevrolet passenger cars, but had Corvette-like body styling in glass-reinforced plastic. Reaction to the Nomad was tremendously favourable, and this convinced Chevrolet to put a Nomad wagon into the Bel Air line for 1955. Through to 1957, 22,898 Nomads were built, along with 10,998 similar Pontiac Safaris. Though this was three times as many cars as Corvettes were produced over the same period, it wasn't enough to meet GM's overheads, and so by 1958 the Nomad and Safari had become conventional estate cars.

Collectors have been searching for the 1954 Motorama cars for a generation, but without success. Rumours continue to persist that the Waldorf Corvette-Nomad was saved by GM and spirited away, despite continued denials by the company. Quite a lot of

For the 1954 Motorama, this Corvette prefigured things to come with glass roll-up windows and a thin-pillar snap-on hardtop. Both features had to wait until 1956 production.

Mounted on the 115in wheelbase Chevrolet passenger car chassis, the Corvette Nomad evolved into the 1955-57 production car, but not as a Corvette. Hardtop-like window styling, sharply raked rear end and curved rear glass were innovative features.

Here's what the airbrush can do! First, an early 1956 model made ready for the shows with a duotoned hardtop and far-ahead narrow-band whitewalls. Next, the same photo retouched by GM for press releases. Neither feature was stock.

Corvette second-generation styling was prefigured in the 1955 Motorama by the LaSalle II, which featured concave scoops in the body sides, not to mention several other less acceptable features.

Motorama specials *do* exist in the GM collection, but the Nomad isn't one of them. Neither does it seem likely that a fortunate collector will find the car buried under a haystack in Podunk, Alabama — but you never can tell.

The 1956 Corvette saw the first restyle since introduction, and many of its components were again derived from GM showcars. The front wings and twin 'power bulges' in the bonnet were copied from the 300SL Gullwing, but the concave body side indentations came directly from the LaSalle II and the Biscayne, which were displayed at the 1955 Motorama. The front wing vents, fitted to the original Waldorf Corvette, were also resurrected for the production 1956 models.

In the summer of 1956, Harlow Curtice was presented with another Corvette special for his personal use. This bright metallic-blue roadster with removable hardtop featured extended front wings cast from the same moulds that had produced the SR-2 racing car. Curtice's version had brushed stainless-steel inserts in the concave sides and genuine knock-off wire wheels. A curious tacked-on tailfin set squarely in the middle of the boot lid mimicked the SR-2. One detail which reached production was a convex bulge at the rear of the body side scoops. This was relocated to the front of the scoop in 1958 production cars.

Also to be seen at the same 1955 Motorama, this LaSalle II saloon attempted to revive the look of the last production LaSalle of 1940, but not at all successfully.

A third styling exercise from 1955, the Chevrolet Biscayne dramatically portrayed a production car shape destined to come for 1958-59.

XP-700, one of Bill Mitchell's first specials after taking over GM Styling upon the retirement of Harley Earl. Far out with bubble hardtop and extended nose, this car did not influence production Corvettes except at the rear. Later it was made into the Shark.

Bill Mitchell's personal SR-2, later shown at the Motorama and raced by Jerry Earl, was a bright red car trimmed in a racing motif, with a huge finned headrest — obviously inspired by the D-type Jaguars — and a small racing windscreen. A metal tonneau cover streamlined the passenger area. Mitchell's SR-2 sported Halibrand knock-off magnesium-alloy wheels shod with racing tyres. The SR-2 had little effect on the shape of future production Corvettes, but it did help establish Mitchell's identity as Harley Earl's successor.

Following his succession to head of GM Styling in 1958, Mitchell created a much wilder dream car for his personal use, the XP-700, intended to display some of his ideas for the future. Thankfully, this weird looking contraption did not greatly affect Mitchell's thinking for production cars. The best lines on it — the flat rear deck and high, rising line off the top of the wheel wells — were adapted for 1961 Corvettes, along with its recessed tail-lights; the huge front overhang and plastic bubble-roof were not.

Also in 1958, Mitchell began work on what was to be the most important Corvette special to date — the first Stingray. He called it a Stingray because he wasn't allowed to call it a Corvette.

Reason? The famous 'ban' on racing was in effect, and Mitchell planned to race the Stingray. Indeed, he'd managed to 'buy' the chassis of the defunct 'mule' test car from Zora Arkus-Duntov's 1956 Sebring programme. The Stingray's open body was actually adapted from a coupe called the Q-car, which Mitchell had wanted to build as the all-new 1960 Corvette — but which had been vetoed by GM management. Bill Mitchell was going to get that Q-car, one way or the other.

The whole concept of the Q-car was aerodynamics — state-of-the-art aero, as we knew it in the 'Fifties. Thus its body had a flat top and a rounded, wind-cheating bottom. 'It was supposed to act in theory like an upside-down airplane wing', one GM designer told the author, 'pressing the tyres to the road for unstickable handling. Actually, we found that the body created lift — the front end would literally leave the road at high speeds. If we only could have gotten it to take off!'

Whatever its aerodynamic failings, the Stingray looked wonderful — a piece of clean-slate thinking at the nadir of Detroit styling, a giant among pygmies. And its taking-off tendencies couldn't have been too great, for Dr Dick Thompson raced the

Dramatic and beautiful, the original Stingray racing car influenced a generation of Corvettes. Shown here it is in its post-racing form, when Mitchell restored it and painted it silver metallic. The headrest was an obvious crib from D-type Jaguars, (and this was not written by an Englishman!).

One idea that didn't come off, the four-seater, stretched-wheelbase Corvette, originally conceived as a possible upstage to the four-seat Thunderbird. Extra length and seats didn't work here any better than they did on the E-type Jaguars. Zora Duntov and Bill Mitchell insisted that a four-seater should *not* be built.

A Mitchell showcar for 1963 sports Bill's favoured outside exhaust system and racing stripes. Mitchell eventually caused the pipes to become standard equipment on certain Stingrays.

Looking just like its namesake, the exotic Shark, designed by Larry Shinoda, was later renamed Mako Shark. Mitchell liked the fade-away paint job because 'monotone dark cars lose lower definition in the shadows — the Shark looked good in any light'.

Several good points to this view of the Mako Shark II... This appears to be the non-running showcar. Cockpit was an electronic wonderland and came complete with organ-style pedals and clumsy-looking fingertip controls under the steering wheel.

Stingray to the C-Modified championship in 1960. This accomplished, Mitchell retired the car from the circuits and refinished it completely. Painted silver metallic, fitted with huge outside exhaust pipes, Halibrand magnesium wheels and a silver leather interior, it became the star of the last Motoramas. Today it remains Bill Mitchell's favourite Corvette — and everyone else's as well. It was the most popular showcar in the history of the Motorama.

The greatest accomplishment of the Stingray occurred not on the circuits, however, but in the boardroom. While Mitchell's Q-car had failed to impress the GM directors, the Stingray succeeded. Suddenly they allowed as how it was a Corvette after all, and furthermore ordained that its styling be the basis of the next generation of production Corvettes. The first production Sting Ray arrived in 1963.

One thing the Sting Ray didn't lead to was a four-seater Corvette. As mentioned earlier, there was a small groundswell among corporate bean-counters to launch a Corvette reply to the highly successful four-seater Ford Thunderbirds. Mitchell's crew rather reluctantly produced a mock-up based around the 1963

Running version of the Mako Shark II, an impressive looking special that directly influenced the design of the 1968 Corvette.

The Manta Ray, of 1969, was an update of the Mako II and made first use of body-coloured polyurethane bumpers front and rear. Six quartz halogens were far beyond USA's stone-age lighting standards, would have been illegal in most states.

Sting Ray coupe and it was photographed alongside a 1962 Thunderbird. Maybe it was the side-by-side comparison that killed it. Zora Duntov said the Corvette was a sports car pure and simple, while the Thunderbird was just a dandified saloon. For whatever reason, the four-seater was never accepted for production.

Next to the Stingray, the most memorable Corvette special has to be the Shark, styled by Larry Shinoda to virtually replicate the fish itself — around which hangs a funny story. Mitchell ordered Shinoda to paint the car to match a mounted and stuffed shark that hung in the lobby of the GM Styling Center. Styling's painters sprayed the car over and over with tiny airbrushes. Each time, Mitchell would look at his stuffed shark and tell them it wasn't right. 'Finally', remembers Ken Eschenbach (head of GM Special Vehicles Division), 'we just stole the damn fish one night and painted *it* to match the car'.

You can fault Mitchell's loud tastes and boisterous self-promotion, perhaps, but you can't fault his instinct for designs that will literally lay people on their ears. The Shark was, and is, a

Recently advertised for $165,000 in the old-car press, the XP-819 was one of the early rear-engined specials and cost GM over $500,000 to build in 1964. The engine was a stock 327; the entire body aft of the cockpit lifted for access.

sensation. 'It has so much animation and so many things are happening on its body', Mitchell said. 'It's really in a class by itself.'

Amazingly, the Shark was based on nothing more radical than the old XP-700 special, itself mainly a restyled 1958 Corvette. It retained the XP-700's clean rear styling and used a 1961 stock chassis. It was built in two months flat, from the first sketches to the finished car, and that includes the time spent trying to match Bill Mitchell's stuffed fish. The basic shape is stock 1963 Sting Ray, but the elongated nose and triple scoops low down behind the gaping 'mouth' remind one of nothing so much as the real fish, one of nature's best examples of streamlining. The Shark was pretty ferocious, too. Its original powerplant was a supercharged 327 V8, but over the years it acquired a 427 engine from a Can-Am racer, rated at 650bhp. In 1965 it was renamed the Mako Shark. The paint scheme has been refined and the interior refinished at least twice. GM still has it.

The Mako Shark II arrived in April 1965. True to GM showcar tradition, it was released to get public reaction to the next generation of production cars, which would arrive with the '68 models. Less exaggerated than the original Shark, it featured a pinched waistline connecting pod-like wings and a gradually sloping fastback with a louvred rear window. In October 1965 a running version of the Mako II was built, with an extended nose and fastback roofline, much cleaner than the original non-running mock-up. Six quartz-halogen headlamps were hidden behind a pop-up panel on the bonnet, and the Mako II was fitted with a 427 ZL-1 V8, plus a cornucopia of special electrical gear including a digital speedometer and a retractable rear bumper.

In 1968 the Mako Shark II was restyled to become the Manta Ray, retaining the shaded duotone paint job, with a new and beautifully styled tail. The goal was to make the boat-tail and horizontal rear wings meet at one point, creating a design which was symmetrical both horizontally and vertically. The Manta Ray also featured a vertical 'sugar scoop' rear light borrowed from production Corvettes, 'off-road' external exhausts, and the first body-coloured urethane bumpers seen on GM products. The bumpers were the only feature of the Manta Ray to work their way into production (in 1973).

XP-819 was the number of a handsome 1964 prototype designed to receive a rear engine, in case Engineering trended in that direction. It didn't, but Chevrolet kept the car around, and it recently surfaced in the old car adverts, priced at $165,000. Hand-built by a team of 11 engineers, the 819 was styled by Larry Shinoda and argued over by Frank Winchell of the Research & Development Group and Zora Arkus-Duntov of the Engineering Group. It used a stock 350 horsepower 327 V8, rear-mounted with a two-speed automatic transaxle, it had rack-and-pinion steering and it weighed 2,600lb. Among Corvette specials, this is happily one of the survivors, but there are far too many of us who want one and not enough cars, or money, to go round. A later evolution of the rear-engine concept came on the Astro II

The Astro-Vette was a study of aerodynamics and had a drag coefficient in the low 3s; the radiator opening was minimal, but adequate. The rollbar was built-in. Finish was white, with a blue interior.

showcar, also by Larry Shinoda, *circa* 1968.

The Astro I, earlier called Astro-Vette, was an ultra-streamlined front-engined special, built from a stock '68 roadster, with a faired-in nose and a tiny oval grille opening. The rear wheels were covered by spats and the rear styling was reminiscent of the Manta Ray. A racing-type screen and side windows flowed into a faired-in, aerodynamic roll-over bar. Painted pearlescent white, the Astro-Vette was never road-tested, but it would have been stable and comfortable at very high speeds.

In 1969 a mild showcar called the Aero Coupe hinted at 1970 production changes, with an egg-crate grille, front wing louvres and flared wheelarches. Later, the Aero Coupe evolved into the extensively modified Mulsanne, used as a pace car for a number of Can-Am races. The Mulsanne has been reworked several times since, and still exists, painted silver metallic, with subtle blue flame trim and housing a full-race LT-1 engine.

The Astro II was another rear-engine evolution, handsomely styled by Larry Shinoda. Big-block V8 would have fitted transversely at rear, and a Wankel was another possibility.

115

The Corvette Mulsanne was derived from the Aero Coupe or Astro-Vette; shown in 1969, it prefigured the stock 1970 egg-crate grille and side louvres.

The most beautiful show Corvette ever created was the Aerovette, which came very close to going into production for 1980. The project began at the New York Auto Show in 1968, where GM displayed the Astro II, or XP-880, a styling exercise created under the direction of Frank Winchell. Back in Detroit, Zora Duntov was working on XP-882, this car specially engineered with production in mind. It had a mid-engine chassis layout which allowed the use of stock Corvette and Oldsmobile Toronado components. Mounted behind the passenger compartment was a big V8 turned 90 degrees to allow the transverse location. The gearbox was located under the forward cylinder bank. A chain ran from the crankshaft to a stock Corvette differential through a short drive shaft that turned a 90-degree corner, lined up with the gearbox, and then passed through the sump encased in a tube.

Chevrolet's new General Manager, John Z. DeLorean, cancelled this project after two XP-882 chassis were built. But after Ford introduced the mid-engine De Tomaso Pantera, DeLorean hauled it out again. Before long it was being used as a test bed for development of GM's then-experimental Wankel rotary engine, first as a two-rotor unit, then with a pair of two-rotor engines combined as a single block. The two-rotor design was handled by Pininfarina, while Bill Mitchell conceived the four-rotor car. Both cars were shown in 1973. But after the Arab oil embargo that year and the advent of the energy crisis, excessive fuel consumption caused the Wankel to be abandoned.

Again in 1977 the four-rotor car was taken off the shelf, fitted with a 400 cid small-block V8. It retained the drive shaft-in-sump arrangement, it was renamed Aerovette and it toured the show circuits for several years, while a lot of people within and without

This XP-880 was the New York showcar for 1968, a non-running prototype. A running version was XP-882, with the same styling and a V8 mid-engine.

GM spoke of production by 1980. The Aerovette had been approved for 1980 by GM Chairman Thomas Murphy — ironically because of the possible threat from none other than John DeLorean and his DMC-12!

We all know what happened to the DMC-12, but what happened to the Aerovette? In late-1977 it was ready for tooling, but by mid-1978 it was on the shelf for good. The simultaneous departure of Bill Mitchell, former Division head and GM President Ed Cole, and Zora Arkus-Duntov, combined to render it 'obsolete'.

The Corvette 2-Rotor was designed by Pininfarina as a test bed for possible Wankel installation. However, the fuel crisis caused it and the Wankel to be dropped.

Sinfully beautiful, the Aerovette came within a hair of production, but the retirement of Mitchell, Duntov and Cole in 1978 finished it.

This Corvette Turbo 3 was fielded in 1981 as a part of turbocharging exercises. Like the new '84, it featured throttle-body fuel injection and an all-aluminium 5.7-litre V8 using heavy-duty heads, block and water pump.

If any car would have been an instant classic the day it was released, that car was the Aerovette. Its styling was almost too perfect — sensational from any angle. In production it would have used the steel platform frame of the XP-882 and Duntov's clever transverse V8, either a 327 or the new 305. Four-speed manual or Turbo Hydramatic transmission would have been offered, and suspension would have been derived from the production Corvette. The 1980 price would have been quite in line at $15,000 to $18,000. It's a shame the Aerovette was never built.

Post-Aerovette showcars have all been mid-engine designs, and there's still a lot to watch at GM in this interesting field. Recent proposals have included a very small Corvette with a turbocharged V6 engine, a mid-engine Berlinetta with styling virtually pirated from the Ferrari 308GTB, Vincent Granatelli's remarkable turbine-powered Corvette (any fuel from kerosene to alcohol, 0-85mph in 5.5 seconds), and finally an Electric. A Corvette *Electric?* Well, if that sounds strange, consider how bizarre the 1984 Corvette would have seemed if you had heard about it in 1964 . . .

CHAPTER 9

The New Generation
1984 to 1988

Every Corvette ever built has been worth collecting; will the latest generation, which debuted as an '84 model (skipping 1983), be more or less collectible than others? It is too early to answer this precisely. Collectibility, where 'Vettes are concerned, does not necessarily vary with the intrinsic merits of the model. The '53 is at once the worst Corvette of all, and the most sought-after model.

It is likely that the new generation will come to be rated by collectors somewhere between the extremes: the 1963-67 Sting Rays and 1957-62 'Fuelies' at the upper end, the 1968-82s at the lower. What we can do right now, however, is review the new car year by year, and give you an idea of what features to expect or avoid if you shop for one.

Completely redesigned, the '84 was 1.1in lower, 8.8in shorter, 2in wider and mounted on a 2in shorter wheelbase. Kerb weight was down by 250lb, and there was more room inside in every direction. The windshield angle was 64 degrees which, with the flush-mounted glass, added a measure of slipperiness. Tested in the Boeing wind-tunnel the '84 returned a Coefficient of Drag of 0.34, and aerodrag was reduced by 23.7 per cent compared with its predecessor.

As for performance, the new car could achieve over 140mph. With the Z51 Performance Handling Package, it would corner at lateral accelerations up to 0.95G and demonstrate impeccable manners during transitional manoeuvres — which was quite a change from certain recent models of the past. The suspension was tuned for firmness and control at speed, 'without cruising-level harshness'. But the quote is from the press releases, and to this writer the '84 was still a very hard-riding car.

Corvette chief engineer Dave McLellan called the new model 'absolutely superior to any production vehicle in its part of the market', without defining exactly what that part was. If he meant luxury two-seaters — on the US market — he wasn't far wrong. To find a peer, he said, 'you have to look at cars produced in extremely limited numbers, and at prices traditionally two or more times that of the Corvette'. (Remember, in America at least, the '84 cost the equivalent of only about £16,000, which must seem a bargain to British readers.)

'It has always been Corvette tradition', Dave McClellan continued, 'to oppose the notion [that] sports cars must be spartan or uncomfortable. With the new design, we continue the tradition.' Accordingly, standard features included air conditioning, electronically tuned AM/FM radio with digital clock and four speakers, power windows, liquid crystal analog display and digital instrumentation, a one-piece removable roof panel, electronic remote-control sports mirrors, ergonomically-designed bucket seats, leather-wrapped steering wheel and shifter boot, an advanced Driver Information System, a security shade to cover the cargo area, a power radio antenna, halogen headlamps and foglamps, and a hatchback controlled by any one of three remote electric releases. There was a built-in theft-deterrent system with starter-interrupt, and care was taken to provide a high degree of safety to the passengers. The seat belts, for example, converted from motion-sensitive to locking at the touch of a button. Chevrolet's concern with comfort and function extended to the cargo area, where the backlight was hinged (as in 1982) for ease of access; and to the new clamshell bonnet, which encompassed the upper halves of the front wings to provide unprecedented accessibility.

The car's styling was the work of Jerry Palmer, who had been

The new 1984 Corvette in European trim, as shown at the Geneva Automobile Show in late 1983. This continental model carried break-away side-view mirrors, metric instrumentation, a rear facia designed for larger licence plates, and special badges and lights, but there was no change in the mechanical specification.

chief designer for Chevrolet Studio Three since 1974, but was enjoying his first ground-up styling assignment. It wasn't an easy one. In a very real sense, the styling of a Corvette is a problem similar to that of Jaguar: you have to make it new, but at the same time retain a distinct, established flavour. Palmer succeeded.

While the traditional Corvette identity has been retained, Palmer said, this required no compromises: 'The car still, for example, has folding headlamps. It has a Corvette "face", even though there are foglamps and park and turn lamps where the air intakes used to be. The front fender vents are still there, as is the large backlight and the functional rear spoiler. The first time people see this car, they're going to know what it is.'

Differences were nevertheless readily apparent. Onlookers didn't believe the car was smaller than the '82, but it was. Palmer commented: 'The new car's massive surfaces, such as the bonnet, are deceiving.'

The '84 began to take shape at Studio Three, in the GM Technical Center at Warren, Michigan, in 1978. But its roots stemmed back to the exciting 4-Rotor showcar of 1972. Since the 4-Rotor was a concept car, Palmer's team experimented with 'a lot of new things — gullwing doors, a windshield angled at the middle in plan view, and a more aerodynamic shape than anything we'd designed before'.

When the rotary engine was abandoned, the 4-Rotor was redesigned for a conventional V8 engine and renamed 'Aerovette', in which form it remained through 1978. Although there were strong rumours that the Aerovette would be the production 1980 model, its impracticalities (doors and windshield, mainly) defeated

The dramatic profile of the '84 shows aerodynamics which led to a 0.34 coefficient of drag. Finely-tuned suspension and P255/50VR – 16 tyres allowed Z51-equipped cars to achieve lateral acceleration of 0.95G.

it. But its basic design philosophy was right — and this provided the starting point for the '84.

Aerodynamics, which by now played a crucial role in car design, was of greater influence on this new car than any previous production Corvette. During wind-tunnel work, the scale model was tested by a sensor passed repeatedly through the wake of the car to record differences between wake pressures and the ambient pressure of the tunnel environment. The result was a detailed picture of the actual pressure variants and vortices created by the passage of the vehicle. This image was more useful than the conventional surface-flow picture, but of course it cost more to develop. Corvette was the first sports car designed with the help of such a tool.

Along with aerodynamics, great emphasis was placed on sheer practicality. 'We wanted to make a car with superior engine-compartment accessibility', Palmer said, 'and we also wanted people to be able to see some of the great hardware — the Cross-Fire injection, air-cleaner, serpentine belt system, electric cooling fan, and the beautiful new aluminium suspension work . . . So we gave it a clamshell bonnet.' These visual pleasures were probably a secondary benefit of the clamshell, whose main purpose was accessibility; but it certainly does allow one to appreciate the intricate and handsome engineering work underneath.

Stylists even had a hand in designing the '84's air cleaner, valve covers, fan shroud, even the sculpted handles of the dipsticks. 'We helped to locate every hose and wire', Palmer continued. 'I remember lengthy discussions concerning the eventual colour of the high-tension cable leading to the spark plugs. We even asked Delco for a new black-and-grey battery so it would go with the rest of the hardware.' The only caveat was that the result had to be utterly functional. It all sounds pretty incredible — and pretty 'Detroit'. But why not? Why shouldn't a fine piece of engineering be beautiful also?

Inside the new Corvette's cockpit, Palmer & Co provided a flat black dash surface under the windscreen, and carefully located controls. The electronic instrument panel was designed for instant reference. In addition to digital readouts on the speedometer and rev-counter, there were large, colourful, easy-to-read LCD fuel analogs. These two instruments and the fuel gauge could be comprehended even by peripheral vision; other critical gauges were

The 1985 model was modified little in design, but Tuned-Port injection V8 gave it another 10bhp and an extra 40lb/ft of torque.

backed by warning lights.

Palmer admitted that, for the first time in a while, designers had 'set aside flashiness and concentrated on basics, like cleanliness, comfort and function. I think this new Corvette is a statement of that principle . . . and it will still appear stylish on the road five or 10 years from now. It's a car designed without compromises'.

Engineering took a back seat to styling on the '84 because some of it was carried over from '82, but many careful alterations occurred, and more would follow in succeeding years. On the '84, for example, an all-new glass hatch and single full-width lift-off roof panel replaced the 1982 T-bar. With the extra body width, the track was increased front and rear. There was more headroom and legroom, and a whopping 6.5in more shoulder room. The new hatch provided nearly 18 cubic feet of cargo space. Yet kerb weight was down to 3,117lb and helped the '84 achieve an EPA mileage rating (Imperial gallons) of 20mpg city, 35mpg highway and 25mpg overall — with automatic transmission. (The 1982 numbers were 18.8, 32.5 and 23.8, respectively.)

Power on the '84 continued to come from the 205bhp 5.7-litre V8 with Cross-Fire fuel injection, electronic spark control and 9:1 compression. Torque was 290lb/ft at 2,800rpm. The engine-transmission combination was rigidly attached to the differential by an aluminium C-section beam. Advantages of this 'backbone' concept included reduced weight and interior packaging gains through the elimination of transmission and differential crossmembers. The standard four-speed manual gearbox was back,

Up front there were virtually no visual differences from the 1984 model. But the price had increased and model-year sales dropped to 39,729.

now with computer-controlled overdrive in three forward gears. The o/d had a hydraulically operated clutch in front and the computer unit at the rear. The computer analyzed several engine functions and factored them into the overdrive for optimum fuel economy in all three overdrive gears; the overdrive was automatically overriden under hard acceleration. Its assembly featured a planetary gearset, which provided an in-series ratio in 2-3-4th gears of 0.67:1. Second-gear overdrive thus gave a final-drive ratio of 1.28:1; third 0.89:1 and top 0.67:1. Direct top had the usual 1:1 ratio.

The '84 was new in its engine accessory-drive system: a double-duty belt, more durable than previous arrangements and requiring less engine power. The cooling system was far more compact and light, and its thermostatically controlled electric fan was mounted directly on the radiator shroud. Bonnet ducts were functional, supplying cool air to the engine compartment. Separate vacuum-modulated doors at each intake regulated airflow, and there was a carburettor hot-air system to assure cold-engine durability. The stainless steel exhaust manifold was redesigned to improve flow.

Entirely new was the welded skeletal steel chassis, a perimeter frame-birdcage unitized structure using high-strength steel to reduce weight in strategic places. The roof, door-sealing panels, dash and sections of the underbody were adhesively bonded to the frame birdcage, and the assembly was completely galvanized and painted. A bolt-on aluminium frame extension supported the rear bumper and the aluminized and painted front suspension crossmember was a bolt-on unit that could be easily replaced. Front and rear bumper facias were bolted on for improved fit and repairability.

The new removable roof panel attached at all four corners. The glass-fibre body benefited from a moulded-in coating process that helped eliminate surface imperfections, with a urethane compound injected into the mould as the panel was formed. Everyone agreed that it was the most perfect Corvette body yet — and certainly the best glass-fibre finish.

The use of so much aluminium in the suspension was a significant technical advance. The familiar short-long-arm independent front suspension used forged aluminium arms and knuckles; as in 1982, conventional coil springs were replaced with the lighter and more durable transverse glass-fibre monoleaf spring. The front suspension had long-life stabilizer link bushings and spindle offset.

An all new five-link independent rear suspension provided wheel control and likewise used much aluminium — for knuckles, control arms and struts. The prop-shaft and drive-shafts were aluminium when the power seat and/or Performance Handling Package were specified. Four-wheel disc brakes with semi-metallic linings were standard on all models, as were 15in cast-aluminium wheels, with 16in wheels available as an option.

The Corvette's wheels, incidentally, were constructed as left-hand and right-hand units (cooling fins were angled appropriately). They should not be swapped laterally. Neither can they be interchanged fore-and-aft on cars fitted with the Performance Handling Package, since in this case they were also designated 'front' and 'rear', and the rear wheels were an inch wider than the front.

How does it all translate into performance? The typical Z51 with automatic would do 0-60mph in just over 7 seconds and the standing-start quarter-mile in 16 seconds at close to 90mph. Spectacular performance indeed — and it was an outstanding handler at all speeds, including triple-digits. The steering response was precise and the chassis was perfectly tuned for fast driving.

That's the up-side, and you can expect it in any '84 Corvette that has been reasonably well maintained. The down-side is threefold, and should be considered: a harsh ride, a considerable amount of cockpit noise, and fuel thirst. The US Government's figures are laboratory calculations and each driver gets different results, but to average over 20 (Imperial) miles per gallon in an '84 Corvette requires the proverbial egg-under-foot technique — or nothing but motorway driving at 55mph, and they don't even do that in America anymore.

The Corvette sold 51,547 examples of the 1984 model, which more than justified GM's high hopes that year. But they haven't maintained that level. The 1985 figure was 39,729, and the total for 1986 was 35,109. This had more to do with accelerating prices than with the cars' merits, because each successive model was an improvement on its predecessor. Nevertheless, it was a shock to Chevrolet Division. After all, the new Corvette had been intended to last until the millenium. It now seems unlikely that this will occur.

Exterior appearance of the 1985 model was hardly changed. The only way it could be told at a glance from the '84 was its straight tailpipes and the legend 'TUNED PORT INDUCTION' on the

The 1986 coupe, again with few exterior changes from previous models, was the first of this generation without a 'handling suspension' option. Prices of coupes are more dependent on condition than model year for the 1984-86 versions.

wings instead of 'CROSS-FIRE INJECTION'. This translated to 230bhp, a 10 per cent improvement over the '84 powerplant, and peak torque of 330lb/ft, up from 290 the year before. At the same time, GM claimed an 11 per cent improvement in fuel economy. The big performance change in the new engine (designated L98 now) was at the top end, where performance was up by about 15 per cent; the car's top speed now met or surpassed the magic 150mph figure.

The box-like plenum of 1984 was replaced by an elongated chamber from which emerged eight curved aluminium runners. Air was ducted from in front of the radiator through a Bosch hot-wire mass airflow sensor and into the plenum, then distributed through the ports and into the combustion chambers. 'Tuned Port Injection' describes a method of increasing the density of air at the intake port by creating a resonant condition synchronous with the valve opening. The tuned runners were designed to help the engine produce more power at lower rpm. It's the same principle reflected in the tall stacks mounted atop many racing engines.

Fuel injectors were mounted in the manifold base plate. A computer calculated precisely how long the injectors were open during each valve event based upon signals it received from the airflow sensor. All '85s were also equipped with a heavy-duty

Acres of glass were a feature of the new-generation cars, an expensive replacement item for collectors; the flip-up hatch, retained from the 1982s, was handy.

cooling package which included a special radiator, high-boost fan and 18psi radiator pressure cap.

The Z51 handling package, retained for '85, now included 9½in wheels in front as well as at the rear, plus Delco-Bilstein shocks and the above-mentioned cooling package. Spring rates on both the standard and Z51 set-ups were reduced from 1984 in an effort to improve ride, while stabilizer bar diameters increased.

More detail changes included a new brake master cylinder incorporating a larger-capacity plastic booster, the first such plastic application in an American car. Manual transmissions were coupled to axles with 8½in ring gears. Complaints about poor instrument readability had resulted in new instrument cluster graphics — bolder and more readable. A leather-trimmed sport seat and automatic temperature control air conditioning were later added to the '85 option list.

Climate control should be avoided if possible. In a car with this

The impressive new 1986 convertible was the first open Corvette since 1975 and Chevrolet sold 7,264 copies. The body was greatly strengthened to improve rigidity.

The '86 at speed, showing the neat covered well for the folding top and the huge rake to the windshield.

Twin turbo Corvettes can be ordered through Chevrolet dealers. New cars are converted by Callaway Cars of Old Lyme, Connecticut, a specialist engine business run by Reeves Callaway. The stock Callaway Corvette (to the right of the picture) was probably the best performance-for-price value in the United States at $58,000 in 1987. In tests run by *Car and Driver* magazine at the Transportation Research Centre in Ohio in summer 1987, a near-stock 1988 version (with 400bhp compared to the standard 382) recorded a top speed of 191.7mph. The white car to the left is Callaway's experimental 'Top Gun' version developing 712bhp. It recorded a phenomenal two-way average of 222.4mph in the *C and D* tests.

Handling, long the bane of past Corvettes, has never been criticized on the newest line. But the author does not recommend the Z51 suspension on pre-1986 models.

small a volume of interior air, it tends to 'search' constantly, alternately heating up and cooling off the occupants. The effect was only slightly alleviated by the increased amount of solar screening in the glass roof panel.

The big news in 1986 was, of course, the return of the convertible — the first open 'Vette in 11 years, which made it good news for Corvette collectors. Not having a soft-top in the line just didn't seem right; after all, of more than 800,000 Corvettes built since 1953, over a quarter were convertibles.

Production of what Chevrolet inaccurately called the 'roadster' began at the Bowling Green, Kentucky assembly plant (sole producer of Corvettes) in mid-January 1986, and cars began arriving at dealerships in February. To celebrate its return, Chevy prepared a bright yellow example to pace the Indianapolis 500 on May 25 — identical to stock except for special track lights. 'I can't think of a higher honour to bestow upon the newest Corvette than to take it off the assembly line and put it on the most famous track in the world', said Chevrolet general manager Robert Burger (with pardonable bias; I always thought Le Mans was more famous).

The '86 convertible is obviously the most collectible Corvette built so far in the post-1982 'New Generation'. It has all the attributes that make the car collector's juices flow — very low production, historic importance and the unbeatable open body style. It was also technically different from the closed models — in effect, a distinct Corvette model. The Tuned-Port V8 developed the same horsepower and torque, but each roadster engine used special lightweight aluminium cylinder heads. These increased the compression ratio from 9.0 to 9.5:1 and shaved 40lb from the engine weight.

Standard transmission was the automatic with fourth-gear overdrive; the four-speed manual with overdrive in the top three gears was a no-cost option. With it, drivers have the joy of seven forward speeds.

All '86 Corvettes featured the Bosch ABS II anti-lock braking system, integrated with the four-wheel disc brakes. A braking development as important today as the first hydraulic brakes in the 1920s, ABS superimposes its logic over a driver's foot, using sophisticated electronics to determine when any wheel is on the verge of locking up. At this point it precisely controls the wheel's braking action to make use of maximum available traction. In

The convertible, in its second year for 1987, saw over 10,000 copies built, the largest number since 1969. Corvette convertibles now account for a third of production, a share not seen since the 1972 model year.

addition, the '86 featured steel-belted Goodyear Eagle P255/50VR-16 radial tyres mounted on 16 × 8½in aluminium-alloy wheels at all four corners.

The conversion job which resulted in the soft-top Corvette was not complicated. The coupe's top could easily be unbolted, though this detracted from the car's stiffness. Reinforcement was therefore added under the glass-fibre bodyshell. The beefing was carried on to the forward frame crossmember, the 'K' braces connecting it to the frame rails, the steeering column and its mounts, and the front torque box. A crossbeam was placed behind the seats, and a strong X-brace under the mid-section. The seatback riser was a double steel panel, and the door latch mechanism was strengthened.

As a result of all this, the convertible proved to be a tighter car than the coupe! 'When body engineer Bill Weaver was asked if these reinforcements could be added to the coupe, he smiled and said it's being looked into', noted *Road & Track*. No joking! The coupe received the same treatment in 1987, and was a much tighter, rattle-free car as a result.

The Corvette top folded traditionally under a special hatch, and was operated manually, a castback to earlier (but far less costly) vintages. 'It is a model of simplicity', argued Chevrolet: 'A three-step procedure takes only seconds. Levers tucked ahead of each sun visor secure the top to the windshield. Turning them outboard releases the top. The same yellow button inside the centre console that controls the hatch release on Corvette coupes controls a glass-fibre panel directly behind the seats on the roadsters. Touch the button once and the top is released, touch it again and the panel pivots upward.' The top, complete with its integrated rear window and velour inner liner, folded neatly into the cavity beneath the panel.

One other 1986 development is worthy of note by the potential collector of late models: the Z51 handling package was no longer available. Some people said that was a shame; this writer says it was no loss, and not likely to be missed on street Corvettes. The Z51 gave slightly better handling at the vast expense of one's backside. The '86 was better off without it, and to me, one of the joys of looking for an '86 is not having to worry whether it's equipped with the handling pack.

One disadvantage must also be mentioned: the new convertible, as simple a conversion as it was, managed to cost the equivalent of £3,000 more than the coupe. Since used car prices have tended to remain high for recent models in good condition, your convertible investment will cost considerably more than a coupe. If you're planning to net a quick profit on a sale two or three years hence, plan again. Chevrolet has been pushing volume up on the convertible, and as long as it is still in production it will be very hard to calculate its ultimate worth as a rarity.

While we are in the process of discussing relative merits, we should record that the 1987-88 Corvette is the best of the breed to date. Buying one, instead of a 1984-86, is going to cost the collector a great deal more money; but it may be worth the money in owner satisfaction.

There was no way to distinguish the '87 from the '86, at least not externally. Even the 'Elizabeth Dole light' was in the same position. (I name this new government widget for its sponsor, the American Secretary of Transportation. It is a little oblong stop light, supplementing the usual ones, which the government requires on the rear of all 1986-and-later cars. It is great idea for avoiding rear-end collisions, until everyone gets used to it, at which point it will be ignored as often as conventional brake lights.)

The '87 changes were under the skin. The 5.7-litre TPI V8 was now equipped with anti-friction, roller-hydraulic lifters and 18-needle roller bearings to support the main roller of each lifter. This translated into slightly improved fuel economy and bhp, but what most drivers noticed was the lower engine noise. The decreased friction will also enhance engine longevity.

Another plus: while a handling package was again available (RPO Z52, for coupes only), it represented an improvement from the old Z51 since it incorporated the base suspension and Bilstein shocks with 13:1 steering gear, 9½in wheels and 3.07 axle.

New also in 1987 was an intriguing low tyre pressure warning system ($325 extra), which may have been worth the money to the high-mileage driver. Wheel modules fastened to the metal rim inside each tyre contained hermetically-sealed pressure sensors and a radio transmitter powered by a long-life power generator. The generator converted the existing mechanical energy in a moving wheel to electrical power — no batteries, no external wires, no connections and no maintenance. The package weighed only 2lb; a precision counterweight was mounted directly opposite each wheel module to preserve tyre balance. If air pressure in any tyre dropped by more than 1psi below a pre-set level, the transmitter activated a dash-mounted warning light. The lower tyre pressure warning

Press pic: GM's Desert Proving Ground in Mesa, Arizona, produces the perfect skid-pad with a recessed sprinkler operating on blue basalt tiles. More than 33,000 bricks form the test road, which is part of a 1¼-mile brake test track.

A major change for '87 was the modified V8 engine with 18-needle roller bearings, which will probably give cars like this coupe extended longevity.

Remarkable for an industry once famed for the annual model change, this 1987 coupe is distinguishable from the original 1984 version only by slightly altered wheels (no crossed flags) and its air dam.

system also included a feature to alert the driver, should a wheel module malfunction occur.

New options for 1987 were a six-way powered passenger seat and a lighted driver's side vanity mirror. The overdrive engaged light and generator warning light were repositioned, to below the tach curve and the centre console, respectively. The four-speed overdrive automatic and four-plus-three manual gearboxes remained available as Corvette transmissions.

Corvettes have long had the dubious distinction of being the 'most-stolen' cars in America. In a continued effort to thwart thieves, 1987 key components carried individual identification. One road test magazine went so far as to wonder if any further theft-precautions were counter-productive, suggesting that thieves look upon each new device as a challenge!

The process of refinement continued as the 1988 models were announced: the '88 was the most powerful new-generation 'Vette yet, the best handler thanks to new ZR-rated tyres, and one of the best stoppers with its ABS. The horsepower boost, from 240 to 245, meant that Corvette power had been raised nearly 20 per cent since 1984. This did not come at any expense in fuel mileage, nor much gain.

Once again, the typical new model was indistinguishable from the year before — except for cars equipped with Option QA1: 9½ × 17in wheels and P275/40ZR-17 Goodyear Eagle tyres. The front suspension was also redesigned to provide zero scrub radius; this reduced brake pull caused by uneven brake forces, a common threat in wet or icy conditions.

What can we say in summary about today's Corvette? Certainly that it is better than yesterday's, and at least technically the best yet. It is one of the world's best two-seat sports cars by any standard of measurement. Few cars can match it off the line, fewer still on those roads where cars may be driven at their limits. 'There may well come the day when the sight of a Corvette in the rear-view mirror of a high-speed *autobahn* cruiser elicits an automatic non-verbalized

The two-model Corvette line-up for 1987.

The thicker wheel spokes that give the '88 a different look are not the only changes to the model. Others include a revised five-link rear suspension and a higher-capacity brake system.

This open-ended view of the '88 engine compartment shows the notable accessibility of today's model.

The 1988 coupe is the fastest Corvette of the present generation, its 0-60mph time claimed to be just over 5 seconds thanks to the new 245bhp Tuned Port Injection V8. Chevrolet expected coupe production to be 25,000 for 1988, if the year went according to plan . . . Zero-scrub suspension has improved braking behaviour on the new model.

response of "Achtung!", followed by a dive for the sanctuary of the right lane', *Road & Track* exclaimed recently. 'When that happens, the lady will have arrived.'

Despite somewhat slower-than-expected sales, the process of refinement will continue for many more years. GM, in more corporate trouble than any, perhaps, since the 1930s, has other problems to occupy it now. There is also less need for change as the Corvette moves up-market, as it very definitely is. The median income of the typical customer today — over $70,000 — converts to about £42,000, and nearly half the buyers have household incomes of $80,000 (£50,000) or more. The median age is now 40, whereas in 1982 it was only 35. About a third of the cars go to executives, salesmen and managers, while lawyers and judges account for another 12 per cent. Intrinsic in this change of character is the type of 'conquest sales' the new Corvettes are making. The number of foreign trade-ins is higher than ever before in Corvette history, and includes quite a number of Lamborghinis, Ferraris and Jaguars.

The next technical breakthrough may well be all-wheel-drive. The twin-turbo Corvette Indy, the latest two-seat mid-engined research vehicle, incorporates this principle, along with all-wheel-steer (steering through the front wheels alone or all four wheels). It also has ABS braking, traction control and active suspension. It was shown for the first time at the Detroit Automobile Show in January 1986. In Corvette experimental showcar history, the Indy is next in line following the transverse-engined Aerovette of 1977.

The Indy takes its name from its engine, a modified version of the dual-overhead-cam Chevy Indy V8 racing engine developed for the 1986 season. Displacement is 2.65 litres, with four valves per cylinder, but the engine is designed to run on petrol rather than the alcohol mix used on Indy race cars. It incorporates a special induction system featuring twin turbochargers and twin air-to-air intercoolers. It is, of course, multi-port fuel-injected.

One purpose of the Indy is to explore the use of advanced electronics. A cathode ray tube atop the centre of the dash, hooked to a rear camera, provides rearward vision. Two other cathode ray tubes, one per door, provide at the flick of a button, data on vehicle dynamics, navigation and engine operation. Each door also houses individual climate and radio controls.

Both seats are secured in a Kevlar tub, the material used in Formula 1 racing cars. The Indy incorporates Kevlar and carbon in a honeycomb-shaped material that offers maximum strength and security.

Rear steering improves low-speed cornering and parking manoeuvrability: direction changes can occur with minimal turning of the front wheels. As a result, wheelhouses may be narrower and less intrusive to the passenger cabin — a big plus, especially on smaller cars. In addition to the packaging benefits, rear-wheel steering promotes improved high-speed handling. On the Indy, it is monitored and controlled by a microprocessor.

Just as in aircraft, where mechanical and hydraulic control links have been replaced with electronic actuation, a computer-controlled system on the Indy uses a sensor which reads throttle pedal position, sending the information to what engineers call the Total Wheel Control System computer, which activates an electric motor to open or close the throttle. This operation interacts with a traction control system to electronically limit wheelspin. The same sensors that read wheel lock-up for the ABS system also send information to a Wheel Control Computer during acceleration. If a wheel accelerates faster than the others, the traction control system sends a signal to the Wheel Control Computer, instructing the system to electronically close or back-off the throttle to deliver maximum useable torque.

Another facet of the Indy's Total Wheel Control System is its new active suspension, eliminating the need for conventional springs, shocks and sway-bars. Wheel and suspension response is optimized electronically for all conditions, maintaining ride quality without compromising handling. Very fast hydraulics and constant input from the Total Wheel Control System change suspension compliance immediately to absorb bumps or to stiffen the vehicle for hard cornering.

Stock Corvettes were being tested with active suspensions similar to the Indy's in early 1986, and the work continues today. While the Indy is not itself intended for production, it serves the purpose of all Corvette specials since Cerv-1: in the words of Chevrolet chief

Digital instrumentation, not to every Corvette owner's liking, is accompanied by well laid-out controls, but the heating-ventilation-air conditioning widgets are set too low.

Not so much a foretaste of a production Corvette to come as a demonstration of advanced technology. The Corvette Indy, a collaborative effort between Lotus and the Italian company Cecomp, brings together such advanced features as four-wheel drive, four-wheel steering, active suspension, ABS braking, a Kevlar/carbon-fibre honeycomb monocoque and electronic instrumentation and control systems. The mid-mounted Ilmore engine is a twin-turbo V8 of CART racing extraction delivering approximately 600bhp. The car was first seen in Detroit early in 1986.

engineer Don Runkle, it 'encourages us to dream a little and to push automobile design and engineering to its outer limit'.

Other modifications, as the new generation moves toward the 1989 and 1990 model years, will be aimed at better aerodynamics and continued weight reduction through the use of lighter materials. A reshaped tail, to cut a few hundredths off the Coefficient of Drag, may show up in a few years. Nobody is saying anything about replacing the V8; it would be like changing the Washington Monument to a pyramid. But a turbocharged V6 is not impossible. It may come out of sheer economics, with broad applications to larger GM automobiles as well. For now, however, there are no V6 plans that anyone will talk about.

We can safely conclude that Corvettes built since 1984 will prove highly collectible. I trust that the foregoing account has explained why. Technically, the newest generation is the most interesting Corvette to come along since the 1963 Sting Ray; maybe even more so.

The Corvette Genève, produced by ASC Inc of Southgate, Michigan, for the 1987 Geneva Show, was suggested by *Car and Driver* magazine to offer an 'accurate early look at the 1990 Corvette', a car which is widely expected to be powered by a 5.7-litre 32-valve all-alloy Lotus V8 producing 400bhp. The 17-inch wheels of the Genève mirror the size adopted for the 1988 production Corvette. Black leather was widely used for the upholstery and interior trim of this show car, along with charcoal suede infill panels.

CHAPTER 10

Buying a Corvette

Model identification and collector preferences

When buying a Corvette, the first rule of thumb is to be sure the car is authentic. It is also the second and the third rule. There's a huge difference in value between a Corvette that is original or authentically restored and one which is equipped with non-authorized options. The following has been compiled by Sam Folz, one of the top Corvette authorities, to recap year-to-year changes and direct you towards the most prized accessories. There are few endemic faults common to Corvettes, but Sam mentions those that exist. You'll probably be surprised to learn that rust IS one of them ...

In the beginning there were only the three Motorama Corvette prototypes. One was built for the 1953 Motorama Show held at the Waldorf-Astoria Hotel in New York City. The others were to be transported in trailers about the United States and Europe so as to whet the appetites of potential buyers — to test the waters for things to come.

The 1953 Corvette, produced in one run of 300 units in the second half of that year, is a close approximation of that show-stopping star of the General Motors exhibit. 'Pure Harley Earl' in design, it is a collector's piece *par excellence*. Unique styling coupled with short supply have elevated today's purchase price. This car, as any early Corvette, should be a good investment if purchased at market.

The early cars are dependable, solid machines. They have good road manners and are no different from their Chevrolet passenger car cousins in providing a high degree of reliability for their loyal owners.

The drawbacks in buying and owning these early examples, aside from a relatively high purchase price, are a very low windscreen that allows a lot of buffeting when the top is down, and a whole series of built-in water leaks when the need to button up occurs. Since there's no way to lock the car, other than the boot, and since parts are very costly to replace, you are not likely to drive one of these to any sort of public area and leave it far from your sight. As all 1953 Corvettes were finished in Polo White with red interior, you have no colour option, should you be lucky enough to find an authentic example.

Comments relating to the 1953 model also apply to 1954, except that now you have a few colour choices. Soft tops were all tan, regardless of the body colour; but Pennant Blue metallic, Sportsman Red, and black bodies were produced. Black is the rarest, blue and red next.

The 1955 models, last of the first generation Corvettes, are very rare and desirable. All the aforementioned liabilities still apply, countered by certain differences worth considering. All but a handful of the 1955s were equipped with the new, hotter 265 cubic inch (4.3-litre) V8, initially coupled only to the two-speed Powerglide automatic transmission. The V8 with its four-barrel carburettor added a lot of oomph to the performance figures, and the weight distribution — having been improved with the lighter engine — afforded lighter steering and better handling. Colours were the predominant white with red interior, plus Gypsy Red with light beige interior, Harvest Gold with yellow, and a few Corvette-Copper jobs with tan interiors.

It cannot be stressed too strongly that authenticity, and complete adherence to factory specifications, is the best assurance of quality in your investment. The car by nature lent itself well to modification of body panels, larger displacement and altered engine types, clutch and transmission swaps, and other changes which detract from value.

SUMMARY:
Points for the purchase of a 1953-1955 Corvette:
1. Styling
2. Scarcity, particularly 1953, 1955 three-speed; 1955 six-cylinder
3. Good road manners
4. Mechanical reliability
5. Quality investment

Points against purchase:
1. Threat of theft of car or parts thereof
2. Lack of choices in engine, transmission options
3. High cost
4. Water leaks in inclement weather

For 1956 the Corvette was further refined in the areas of comfort, style and power. In addition to white, body colours were offered in black, blue, red, copper and green. The old Blue Flame Six was dropped and the 265 cid V8 with one or two (progressive linkage) four-barrel carburettors was teamed with a now-very-available three-speed manual transmission or the two-speed Powerglide automatic. The windshield was increased in height about 1-1½ inches and the awkward soft-top mechanism was vastly improved. A power-top option countered sales resistance to the earlier, hard-to-erect canvas.

Other improvements and options included an in-production change from the old recirculating heater from 1953 to a true outside-air type, greatly improving demisting; roll-up side windows with electric power option; auxiliary glass-fibre hardtop; and transistorized radio. For the first time, the Corvette became a convertible, as opposed to a roadster. With outside door handles and locks, the cockpit could now be made secure from theft. A locking glove box was also included between the two bucket seats.

Another styling feature that was new to the 1956 Corvette was the 'waffle'-pattern vinyl on seats, sidewall and (if ordered) headlining of the optional hardtop. This material was first seen on the 1954 'Corvair' prototype which was a 1954 Corvette modified to fastback configuration. It later appeared on the 1955 Chevrolet Nomad station wagon.

Externally, styling was noticeably changed by raising the headlamps and projecting them forward, by adding a streamlined cove or recess from the front wheelarch back long the door, and by dropping the tail-lamp pods and recessing the lamps into the resultant voids. The aforementioned cove also provided a unique scheme for two-toning the colours and the cove would be silver on black, white, or blue bodies, and light beige on all others.

The 1957 models were only slightly revised in style from the 1956s. The big news was under the bonnet with the now famous small-block 283 V8, providing increases in torque and horsepower. Further, Rochester Products' Ramjet fuel injection became available as a regular production option, and in top tune was rated at 283 horsepower, or one bhp per cubic inch of displacement. Of the 6,339 1957 models produced, only about 15 per cent were ordered with fuel injection. These were identified with special nameplates on wings and bootlid. One version, built in extremely limited numbers, allowed cool air to enter the induction system through an air box on the inner left skirt, and this built-for-racing 'fuelie' is the collector's 'top of the line' for this series.

Colours were continued from 1956, with a silver option added to the list. When two-toned, this body carried a white cove. One other significant mechanical change for 1957 was the introduction of a four-speed transmission. Early cars did not have this option available, so even with the fuel injection unit specified, cars built prior to May 1, 1957 had either three-speed manual or Powerglide automatic gearboxes.

In summary, 1957 models were refined, superior performing cars when compared to their predecessors, and are thus most desirable. Particularly desirable are those with fuel injection and four-speed gearboxes.

SUMMARY:
Points for the purchase of a 1956-1957 Corvette:
1. Styling
2. Weatherproof windows and top, plus hardtop option
3. Increased acceleration and top speed
4. Performance options in 1957, including fuel injection, wider-based wheels, four-speed transmission and performance suspension
5. Quality investment

6. Variety of colours

Points against purchase:
   1. High purchase cost
   2. Inability of stock-size tyres and brakes to cope with the new performance level, particularly fuel-injected 1957 models

For 1958, the basic 1957 Corvette chassis with the 283 cid V8 (4.7-litre) with its options of three- or four-speed manual transmission or two-speed Powerglide automatic were continued. The big news was a major styling revision. All GM cars, Corvette included, received the quad-headlamp treatment. The large grille oval was replaced with a smaller one to be flanked by fake outer grilles under the headlamp group.

The bonnet received a fake set of louvres between the raised wind-splits, just ahead of the scuttle ventilator, and the boot received a pair of heavy chrome bars from the folding top compartment rearwards to the rear bumper. The cockpit was totally restyled with a new instrument cluster in full view of the driver and a central console for heater, radio and clock. The vinyl trim for 1958 was to be a one-model-year specification with a pronounced 'pebble' grain, not seen before or since in Corvettes.

For 1959, the 1958 model was 'cleaned up' by elimination of the fake hood louvres, the 'wild' grain in the vinyl, and the heavy chrome trunk bars. No serious mechanical or power alterations were made from 1958 specifications, but colours were added such as black (which replaced the unique Charcoal of 1958).

In 1960, the emphasis was once again on power and performance over the 1959 model. The top fuel injection option gave 315 horsepower from the rugged 4.8-litre V8 and the quality of these cars was at a new high. Nineteen-sixty was really a vintage year for Corvette. This model can be identified most easily from its 1959 brother by noting that seams in the seats run fore-and-aft, rather than across. Also, small red and blue bars were added to the trim panel in the cove ahead of the passenger's seat.

SUMMARY:
Points for the purchase of a 1958-1960 Corvette:
   1. Style, the 1958 model being the most distinctive
   2. Performance, the 1960 model leading in this respect

   3. Comfort and reliability
   4. Quality investment

Points against purchase:
   1. High cost
   2. Stock tyre size (6.70 × 15) too small for high-performance engine options

The 1961 Corvette gave just a glimpse of things to come when the upturned or so-called 'duck butt' rear end from the Stingray race car was adapted to the 1960 body style. The remaining grille teeth from the 1953 model were removed and replaced with a slightly convex anodized aluminium screen. For the first time, the dual exhausts did not exit through the body or corner trim at the rear, but terminated with pipes at the side below the rear wings, just behind the wheels. With the new rear design, tail-lamps were changed from single to dual round style and the lid for the boot became top-opening due to the upswept rear. Mechanically, the 1961 was not very different from the previous year, with its power rating unchanged.

In 1962, the same styling was continued, but with a cleaner, leaner look. The anodized grille screen went from bright to matt-black. The bright trim around the side depression was dropped in favour of a raised lip in the glass-fibre body, the contrast colour option was deleted for the first time since 1953 and wide whitewall tyres were not installed. Instead, the new white stripe tyre (optional) was added to the low look.

The further big news for 1962 was the enlargement of the 283 engine to 327 cubic inches (5.4 litres) with significant increases in torque and horsepower. The 1962 Corvette is a real screamer and when driven in a top state of tune the 6.70 × 15 tyres chirp on dry tarmac when an inexperienced driver changes up from first to second and from second to third, so brutal is the available torque. Many believe the 1962 Corvette to be the cream of the first-generation straight-axle Corvettes, and market prices reflect that feeling.

SUMMARY:
Points for the purchase of a 1961-1962 Corvette:
   1. Styling (see below)
   2. Performance, with emphasis on 1962 340 and 360 horsepower

3. Rugged, reliable and well constructed
4. Quality investment

Points against purchase:
1. Styling, if you don't care for the 'old' front with the 'new' rear
2. High cost
3. Tricky in the hands of a novice

It should be noted that when considering the purchase of *any* 1953-1962 Corvette, *the frame must be examined carefully for corrosion*. Because these cars have 'rust-free' bodies, people tend to assume that the entire car resists the dreaded tinworm. Not necessarily so. Make sure that the side rails are sound, and pay particular attention to the rear crossmember. This component carries the aft-end spring shackles, and having a solid part is critical. (The rear crossmember is now being reproduced.)

The 'hottest' area in the Corvette market today begins with the 1963 Sting Ray, continuing through to the 1967, and final, version of this 'mid-year' series.

In 1963, engines continued in 327 cubic inch (5.4-litre) form from the previous year with the same horsepower ratings. Three- and optional four-speed manual transmissions were continued and were most popular, while the two-speed Powerglide automatic in its new-from-1962 aluminium case, satisfied a smaller number of shiftless devotees. Styling and engineering made big news. The now-famous split-window coupe with its boat-tail fastback was, and still is, a sensation. Convertible coupes, with or without the lift-off hardtop option, were continued, and proved to be just as good sellers as the fastback, each taking 50 per cent of 1963 production.

The new Corvette performed very well and with a new, longer list of options it sold well for street, drag strip, or road course use. New colours included for the first time Sebring Silver, which cost extra due to the controlled uniform size of the metallic particles. It still is a popular option among collectors, and such '62s are worth a small premium.

Nineteen-sixty-four Corvettes remained strong in the market and were nicely refined. They continued with only subtle changes and the convertible moved ahead in sales at 63 per cent of the total. The controversial divided rear window was removed from coupes, to the delight of Engineering and the dismay of Styling.

Also gone were the fake bonnet vents used at the front of the 1963. New were genuine knock-off aluminium wheels, which had been announced but not supplied for the prior year.

A small working cabin blower was installed in the left sail panel of the coupe, just behind the door, to improve air flow in a closed car. A matching but non-functional grille was added to the right side. Horsepower was increased slightly at the top end of the option list, but the famous RPO Z-06 of 1963 was dropped due to GM's ban on factory racers. The 36-gallon (US) fuel tank — option N-03 — was continued, however. Tyre size for 1964 was changed for the first time since 1953 to 7.75 × 15 from the former 6.70. Air conditioning, which found its way into a few cars very late in 1963, was now an available and desirable 1964 option.

The optional radio for Corvette (mandatory in 1953 until late 1955) had always been a signal-seeking AM type until early in 1963, when a running change was made to an AM-FM unit. These dual-band radios did not have the so-called 'Wonderbar', or signal-seeking feature.

SUMMARY:
Points for the purchase of a 1963-1964 Corvette:
1. Styling; the split-window 1963 coupes are especially desirable
2. Performance
3. Handling
4. A quality investment, particularly with Z-06 option in 1963, fuel injection, air conditioning and leather seats

Points against purchase:
1. May have too firm a ride for street use, should you find one with competition brakes and suspension
2. High cost

Particular attention is paid by collectors to the 1965 model, the first year for four-wheel disc brakes. Some 1965 models carried four-wheel drum-types at the buyer's option, while parts remained in the inventory; afterwards all cars were supplied with discs. Also in 1965, the big-block 396 cid (6.5-litre) V8 became available as an option, rated at 425 instead of 375 horsepower for the final small-block 'fuelie'.

The 1963-1967 Corvette coupes are led in value by the famous

1963 'split window'. Collectors place about a $3,000-4,000 premium on that model when in concours condition, as compared to 1964-1967 models. When the car is in lesser condition, however, the premium average for the 1963 coupe is reduced to $1,000-2,000.

The five completed Grand Sport racing versions of the 1963 model are all alive and accounted for and have values in excess of $100,000. Convertibles carry premium prices for 1963 as the first of a series, but in this case $1,000-1,500 would represent the average amount.

Absolute top-value cars must carry factory-installed air conditioning, very rare for 1963. This feature adds $500-1,000 to the typical car's value. Optional aluminium knock-off wheels add about $1,000 for originals, although good reproductions are now available. The leather interior is also highly prized, as is the L-88, high-performance V8, which could add as much as $1,000 to 1967 models. In any of the 1963-1965 Corvettes, fuel injection can add as much as $2,000 to collector value and, along with air conditioning, heads the desired option list.

In 1966, without a fuel injected option, the available big-block engine was opened up to 427 cubic inches (7 litres) in 390 and 425 horsepower ratings, and cars so equipped received a new, special bonnet design so that they were instantly recognizable.

Both 1965 and 1966 Corvettes had a special teak steering wheel, with or without a telescopic steering column as optional equipment. Both features are highly prized today.

As with all collector Corvettes, one should pay special attention to the fully optioned cars. That would include fuel injection, a hardtop for convertibles, aluminium wheels, air conditioning, leather seats, power windows and (where offered) the teak wheel with telescoping column. In 1966 and 1967, particularly, attention should be paid to cars with the 427 cid engine, although the 327 is far more tractable. A high performance, high compression engine, the 427 can be adapted to run well on today's lower octane fuels.

The styling of the 1967 car appeals to many because of the unique, functional front wing vent, the cleaner body *sans* emblems, and the fact that it's the finale for this beautiful series. Federal regulations came into being after these cars were designed, but dropping the teak wheel and removing wheel-hub knock-offs were the only concessions that had to be made in 1967.

SUMMARY:
Points for the purchase of a 1965-1967 Corvette:
1. Styling
2. Performance
3. Sound investment with air-conditioned coupes, 1965 fuel injection and 1967 models leading the list

Points against purchase:
1. High cost
2. Unavailability of proper fuel for top compression engines
3. Harsh ride with optional suspension

In 1968 a whole new body and image appeared on the completely altered Corvette. The chassis and suspension were essentially unchanged from the 1963-67 series, although the road wheels were increased in width from the 6in of 1967 to 7in for 1968 and 8in for 1969 and later models. The Corvette finally was able to put a large enough footprint on the road to handle the performance potential of the big engines.

The styling for the new car was slightly modified and not quite as daring as the Mako Shark showcar from which it had been derived. The lower roof line and more severely raked windscreen meant that the passengers came closer to 'wearing' this design, rather than sitting in it as before. It was more stable aerodynamically than the preceding series, with less lift at front and rear at speed, but visibility from the cockpit was not quite as good due to the low seating and prominent humps that provided wheel clearance.

Nineteen-sixty-eight saw five optional 427 cubic inch engines retained, offering horsepower ratings from 390 to 435. Along with this power were three optional four-speed manual transmissions and, for the first time, the rugged three-speed Hydramatic was used in place of the old two-speed Powerglide when an automatic was specified.

The 1969 units were similar to the 1968s, with Stingray emblems reappearing above side vents, and with redesigned outside door release mechanisms. In the engine compartment a small-block 350 cid (5.7-litre) appeared for what would be a long reign, replacing the venerable 327.

In 1970 came minor refinements, with the front wing vent redesigned from four vertical slots to an egg-crate design that would carry through to 1972. The optional 427 cid engines were opened up to 454 cid (7.5 litres), the largest ever to be offered in a Corvette, and available through to the 1974 model.

Nineteen-seventy-one saw a twin of 1970, with a continuation of styling details. For the first time, all Corvette engines were designed to run satisfactorily on no-lead or low-lead fuel, as well as the usual leaded regular, having an octane rating of 91 or better.

The 1972 was the last model in the new series to have a chrome front bumper, an 'egg box' grille, and the Kamm back with chromed bumpers. It was the first of a long line to include a built-in anti-theft alarm system. 1972 was also the last year when the LT-1 high-revving, small-block (350 cubic inch) engine with solid valve lifters could be ordered.

In 1973 the front bumper was altered to meet the new Federal 5mph standard, and its soft nose in body colour gave the Corvette a new face. The rear treatment was carried over for the last time, making the 1973 model unique. The side vent on the front wing was opened up to one port without grille cover. Engine options were reduced to one 350 and one 454, both with hydraulic valve lifters. The cast aluminium wheel option announced and pictured for this car was cancelled, and the few that were actually shipped were recalled, not to appear again until 1976.

The 1974 models received the new 5mph rear bumper assembly. This first effort was easily identifiable from the rear by the two-piece cover, a one-model-year characteristic. This was also the final year for a real dual exhaust system, and the last year that leaded fuel could be used legally. It was also the last appearance for the 454 engine.

The last of the convertibles was produced in 1975. The Corvette, along with all other GM passenger cars, now received the new catalytic converter, and only one optional engine was offered — the 205 (net) horsepower L-82.

Nineteen-seventy-six found only minor trim and identification changes. Underneath, the car received a steel belly pan at the front to shield against heat from the warmer-running engine and heat from the converter. The long awaited aluminium wheels were finally available. Late-1976 cars were all assembled with the formerly optional interior, including leather seats, as the race-ready Corvettes of the past were no more and Grand Touring became the Corvette's *forte*.

The 1977 models brought minor trim changes such as the removal of the bright trim at the windscreen posts and a new nose emblem. Sport mirrors with inside control were a popular option, along with stereo radio and air conditioning. A new luggage rack and hardware now allowed the removable roof panels to be carried outside, leaving additional luggage space inside.

The 1978 model marked the Corvette's 25th Anniversary Year, and special front and rear emblems appeared to commemorate the event. The small, flat, rear window within the sail panels was replaced with a large formed piece giving a semi-fastback effect that continued through to the 1982 model.

Two special models are of interest to collectors: The Silver/Charcoal 'Silver Anniversary' option, with its striped sport mirrors, aluminium wheels and other trim, was sold as option B2Z — Silver Anniversary Paint; the other was the infamous 'Pace Car' replica, built in one run of 6,500, or just over one per dealer. These cars were finished in black over silver with red accents, and Indianapolis Speedway decals, the latter supplied loose for owner installation. Interior trim was altered for these cars, and the most notable feature was the 'new' bucket seat, replacing those seen so many years before. This seat then became the standard in 1979 and later.

The 1980 units appeared 'newer' by using altered front and rear bumper caps with integrated spoilers, and revised side vents on the front wings. The new Federal 85mph speedometer regrettably appeared, and former options such as air conditioning and the tilt/telescope steering column became standard as the Corvette moved still further away from the performance to the boulevardier image.

The emphasis for 1981 and 1982 Corvettes was weight reduction on otherwise scarcely altered cars. A new glass-fibre rear spring replaced the older metal one, for example, with good results.

As per 1953-1962 models, corrosion of frame components should not be overlooked when purchase of a 1968-1982 model is contemplated. The 1963-1982 chassis, with its independent rear suspension, should be carefully examined, paying particular attention to the main frame rails just forward of the kick-up at the rear axle. Heed should be paid to the metal transverse rear spring on 1980 and other versions.

SUMMARY:
Points for the purchase of a 1968-1982 Corvette:
1. Styling — if you like it
2. Potential price appreciation as new replacement costs skyrocket, especially the 1975 convertible
3. Performance, particularly early L-79 and L-88 engine option versions, also LT-1 of 1970-1972
4. Comfortable Grand Tourer
5. Modest pricing compared with 1967 and earlier models

Points against purchase:
1. Low seating and wheel house rise that affects visibility
2. High fuel consumption/cost/availability for big block and high performance versions

In the 1968-1982 Corvette Stingray group, there is no clear-cut popularity winner among collectors, although the last-of-a-type 1975 convertible can clearly be seen to have escalating interest and price appreciation. Both coupes and convertibles of 1968 as the first-of-a-type show collector strength. But there were some serious glass-fibre panel problems with this model, so condition of body and finish are critical to the car's value.

As with the 1963-1967 series, the fully-optioned cars catch the collector's eye. Try to get factory air conditioning, leather interior and AM-FM stereo. Tilt and telescope steering column, power steering and power brakes are also highly favoured.

Big-block engine options attract certain collectors, but the lack of high-octane fuel makes the small-block engines easier to live with in the USA, though this is not yet a problem in Europe.

From 1976 on, the aluminium road wheel option is also sought after. Note that when that option was specified, you received only four, the spare being plain steel.

The new line of cars which began in mid-1983 as a 1984 model has now been around long enough to allow us to make a few considered judgments about their relative merits. Broadly speaking, when it comes to this generation, the latest model is usually the best, and convertibles stand to be considerably more sought-after than coupes at the current rate of production. (Note, however, that for 1987 the convertible reached a 1:2 ratio with the coupe; its increasing share of the total production bears watching.)

The general merits of the '84 are extensively discussed in Chapter 9. As the first of the new series, does it stand to be the most collectible? The answer is probably 'no'.

There is, first and foremost, the matter of production: the '84 saw the highest volume of all four model years complete at this writing, and no '84 model was a convertible. To get the obvious point out of the way, the 1986 and later Corvette convertibles are about twice as desirable (if not yet priced at twice the cost of) their coupe counterparts. This is simply because of collectors' longstanding love affair with open cars, and because of the convertible's relatively low production.

Turning now to technical merits, we discern steady improvements in the breed over the past five years. The 1985 model with its Tuned Port — instead of Cross Fire — injection had modestly more horsepower and considerably more torque, but the changes will generally be experienced at the top end of the performance curve: not something you are in, most of the time.

Aside from the arrival of the ultra-desirable convertible, we may count the absence of optional Z51 suspension as a 'plus' on the '86 model. That kidney-whipping ride may help somewhat on a slalom course, but it is a dreadful thing to live with; if you shop for an '86, you will not have to worry about running into it.

Another option devoutly to be avoided on '86 and later models is Climate Control, which is a contradiction in terms in a cockpit as small as the Corvette's, and which tends to 'search' back and forth, alternately spraying you with Arctic chill and tropical heat. Avoid it if you can.

The '87 was clearly the best of the new breed yet. The new V8 with its anti-friction hydraulic lifters and 18-needle roller bearings is quieter, and stands to last longer, than any of its predecessors. The new Z52 handling suspension option was a major improvement on the 1984-85 Z51, and well worth having. (It consisted of Bilstein shocks, 13:1 ratio steering, 9½in wheels and a 3.07 automatic axle ratio.)

The '88 is virtually a repeat of the '87, but it is worth noting that it came with an additional 5 horsepower (245 net, a 20 per cent gain from the '84 spec), and a small but useful modification to the front suspension: zero scrub radius translates to reduced chance of brake pull on hard deceleration. So the '88 is thus far the most desirable Corvette — except for its price, which at this writing is 'full sticker and then some' for convertible models in some 'Vette-short markets.

Obviously all these cars are too new to evaluate finally as to their collectibility. Just as obviously, in the current used car line-ups, they vary in price, although not as much as you might think: a 20,000-mile '84 will certainly cost you more than a 80,000-mile '86, and yes, if that's the major difference, you should buy the '84.

SUMMARY:
Points for the purchase of a 1984-1988 Corvette:
1. The most technically advanced Corvette in history.
2. State-of-the-art styling and engineering.
3. History to date on the used car market indicates very little depreciation if you take care of it.
4. The most luxurious and comfortable of any Corvette generation.
5. Generally good quality of fit and finish, and far better than the previous 1968-82 models.

Points against purchase:
1. High price. Base prices jumped by a huge amount for the '84 compared to the '82 and have been rising every year since; used car values show little depreciation, which is good, except for buyers.
2. A definite change in character from the old, rorty screamers of the 'Fifties and 'Sixties, which may, in the long run make the current generation a much less interesting car to collectors of the next century.

CHAPTER 11

Corvette ownership

Clubs, spares and restoration

National Corvette Restorers Society
Any account of the spare parts and restoration scene, where Corvettes are concerned, has to begin with the National Corvette Restorers Society (NCRS) – one of the best organized single-make clubs in the world. Initially, NCRS' interest was in pre-1956 models, but it was later expanded to cover all cars through to 1977. No doubt it will eventually move on into later territory, as the Corvette design of 1978-82 is now passing into history.

Remarkably, this 13,000-member car club is only two decades old, having been founded when seven early-model Corvette owners met at the Holiday Inn, Angola, Indiana, for the purpose of organization. The time had come, they felt, because very little original information on the earlier models was held by Chevrolet, and burgeoning interest in the marque was creating a demand that Chevrolet Division itself could not fill. Today, NCRS is represented in all 50 states and a dozen foreign countries.

Number-one service of NCRS is a quarterly magazine, *The Corvette Restorer*, which began as a 16-page newsletter but is today published in offset 40-50-page format and has a crack staff of devoted writers who cover every aspect of Corvette information. In between copies of the *Restorer* comes a bimonthly *NCRS Driveline* newsletter, filled with advertisements and news of Corvette happenings worldwide.

NCRS holds at least one convention annually, which includes one of the more spectacular concours judging events in the classic or vintage car field. Parts and literature swaps, technical sessions and question-and-answer seminars are held along with the concours. Corvette personalities who have been guests at NCRS banquets include author Karl Ludvigsen, Molded Fiber Glass Company's Robert Morrison, prominent Corvette publisher Mike Antonick, former chief Corvette engineer Zora Arkus-Duntov and chief engineer Dave McLellan.

NCRS makes a point about helping Corvette owners at its *concours d'elegance*, which, they say, 'is not a contest to see which car is the cleanest; rather it's an event that gives owners a chance to have other knowledgeable enthusiasts discuss possible areas of vehicle improvement, as well as note areas of excellence and originality. In fact, to discourage competition among owners, a "flight system" is used, and ribbons are presented to any and all cars achieving a minimum of 70 out of 100 points in scoring. Also, mileage points are offered as an incentive to owners who drive their Corvettes to the convention'.

To join the National Corvette Restorers Society, send $30 (United States), $35 (Canada), or $45 (everywhere else – publications are sent airmail), together with your name, address, telephone number, year and serial number of all Corvettes owned, to NCRS Inc, 6291 Day Road, Cincinnati, Ohio 45252, USA, tel: (513) 385-8526, fax: (513) 585-8554.

Other Corvette clubs
The National Corvette Owners Association, 900 South Washington Street, Falls Church, Virginia 22046, USA,

claims 10,000 members and publishes *For Vettes Only*. Subscription is $32 per year.

The National Council of Corvette Clubs Inc is open to Corvette owners, with subscriptions of $30 (initial) and $20 (renewal). The Council's address is PO Box 5032, Lafayette, Indiana 47903, USA.

Independent regional Corvette clubs include:
Cascade Corvette Club, PO Box 363, Eugene, Oregon 97440, USA.
Corvettes Unlimited, 1120 Fairmont Avenue, Vineland, New Jersey 08360, USA.

Specialized Corvette publications include:
Vette Vues. Published monthly; about 80 per cent advertising for cars, parts, literature, models and services. *Vette Vues* Enterprises Inc, PO Box 76270, Atlanta, Georgia 30358, USA.

Corvette spare parts and services

4-Speeds by Darrell, PO Box 110, 3 Water Street, Vermilion, IL 51955, USA. Tel: (217) 275-3743; fax: (217) 275-3515. Transmissions.
AM Racing Inc, PO Box 451, Danvers, MA 01923, USA. Tel/fax: (508) 774-4613. Race prep, sales and vintage racing.
American Restorations Unlimited, 14 Meakin Avenue, PO Box 34, Rochelle Park, NJ 07662, USA. Tel: (201) 843-3567; fax: (201) 843-3238. Restoration parts.
Andover Automotive Inc, PO Box 3143, Laurel, MD 20709, USA. Tel: (410) 381-6700; fax: (410) 381-5703. Parts, seat belts.
Auto Accessories of America Inc, PO Box 427, Route 322, Bealsburg, PA 16827, USA. Tel: (800) 458-3475; fax: (814) 364-9615; foreign: (814) 364-2141. Accessories, glassfibre, interiors, parts.
B & B Cylinder Head, 320 Washington Street, West Warwick, RI 02893, USA. Tel: (401) 828-4900. Cylinder heads.
Bairs Corvettes, 316 Franklin Street, Linesville, PA 16424, USA. Tel: (814) 683-4223; fax: (814) 683-5200. Parts, service.
Balfour, 15 John Dietsch Boulevard, North Attleboro, MA 02763, USA. Tel: (508) 699-6500; fax: (508) 699-0297. Jewellery.
Beaky's Corvette Corner, RR8, London, ONT N6A 4C3, Canada. Tel: (519) 266-3365; fax: (519) 268-3373. Parts.
Beard Auto Machine, 631 Pine Avenue, Albany, GA 31702, USA. Tel: (912) 436-8888. Restoration, engines.

Bud's Chevrolets, Corvettes, ZR1s, PO Box 128, St Mary's, OH 45885, USA. Tel: (800) 688-2837; fax: (419) 304-4781. Cars.
Sal Carbone's Restoration Parts, Outlet, 55 Clark Road, Bolton, CT 06043, USA. Tel: (203) 649-2241. Parts, smog systems.
CBS Performance Automotive, 1724 Armstrong Avenue, Colorado Springs, CO 80904, USA. Tel: (800) 685-1492; fax: (719) 578-9485. Ignition systems, performance products.
Chevrolet Parts Obsolete, PO Box 10137, Santa Ana, CA 92711-0137, USA. Tel: (909) 279-2633; fax: (719) 279-4013. Accessories, parts.
Chicago Corvette Supply, 7322 S Archer Road, Justice, IL 00458, USA. Tel: (708) 458-2500; fax: (708) 458-2662. Parts.
Classic Profiles Inc, 5770 W Kinnickinnic River Parkway, West Allis, WI 53219, USA. Tel: (414) 328-9866; fax: (414) 328-1906. Artwork.
Collector's Helper, 216 Route 17N, Upper Saddle River, NJ 07458, USA. Tel: (201) 327-8904. Brakes, parts, restoration.
Conte's Corvettes & Classics, 851 W Wheat Road, Vineland, NJ 08350, USA. Tel: (609) 692-0087; fax: (609) 692-1009. Leasing, parts, sales, service.
Corvette & High-Performance, Division of Classic & High Performance Inc, 2840 Black Lake Boulevard SW #D, Olympia, WA 98512, USA. Tel: (360) 754-7890. Accessories, parts.
Corvette Central, 5865 Sawyer Road, Dept HM, Sawyer, MI 49125, USA. Tel: (616) 426-3342; fax: (616) 426-4108. Accessories, parts.
Corvette Enterprise Brokerage, The Power Broker, 52 Van Houten Avenue, Passaic Park, NJ 07055, USA. Tel: (201) 472-7021. Appraiser, broker, car locator, investment planning.
Corvette Rubber Company, 10640 W Cadillac Road, Cadillac, MI 49601, USA. Tel: (616) 779-2888; fax: (616) 779-9833. Rubber products, weatherstripping.
Corvette Specialists of Maryland Inc, 1912 Liberty Road, Eldersburg, MD 21784, USA. Tel: (410) 795-3180; (410) 795-3247. Parts, restoration, service.
Corvette World, RD 9, Box 770, Dept H, Greensburg, PA 15601, USA. Tel: (412) 837-8600; fax: (412) 837-4420. Accessories, parts.
Davies Corvette, 5130 Main Street, New Port Richey, FL 34653, USA. Tel: (813) 842-8000; fax: (800) 236-2383. Customizing, parts, restoration.
Dean's Wiper Transmission Service, Dean Andrew Rehse, 16367 Martincott Road, Powny, CA 92064, USA. Tel: (619) 451-1933; fax: (619) 451-1999. Wiper transmission service.
Doug's Corvette Service, 11634 Vanowen Street, North Hollywood, CA 91605, USA. Tel: (518) 765-9117. Race prep, repairs.
Dr Rebuild, PO Box 6253, Bridgeport, CT 06606, USA. Tel: (203) 366-1332. Parts.
Dr Vette, 212 7th Street SW, New Philadelphia, OH 44653, USA. Tel: (216) 339-3370, orders (800) 878-1022; fax: (216) 339-6640. Brakes, fuel system parts, repairs.

Eagle Products, 4425 Edgewood Road, Louisville, TN 37777, USA. Tel: (800) 977-0030; fax: (815) 977-1024. Wire harnesses.

EC Products Design Inc, PO Box 2380, Atascadero, CA 93423, USA. Tel: (800) 488-5209; fax: (805) 466-4782. Accessories, parts.

Eckler's Quality Parts & Accessories for Corvettes, PO Box 5637, Titusville, FL 32783, USA. Tel: (800) 327-48868; fax: (407) 383-2059. Accessories, parts.

Howard's Corvettes Inc, RR 3, Box 102, Sioux Falls, SD 57106, USA. Tel: (605) 368-5233. Parts.

J & D Corvette, 9833 Alendra Boulevard, Bellflower, CA 90706, USA. Tel: (800-VETTE-JD) 838-8353; fax: (310) 804-5210. Parts.

Joe's Chevy Parts, PO Box 124, Lindenhurst, NY 11757, USA. Tel: (516) 226-9169. Used parts.

Rudy R Koch, PO Box 291, Chester Heights, PA 19017, USA. Tel/fax: (610) 459-8721. Manuals.

Lectric Limited Inc, 7322 S Archer Road, Justice, IL 60458, USA. Tel: (708) 563-0400; fax: (708) 458-2662. Parts.

Long Island Corvette Supply Inc, 1445 Strong Avenue, Copingue, NY 11726-3227, USA. Tel: (516) 225-3000; fax: (516) 225-5030. Parts.

Marcel's Corvette Parts, 15100 Lee Road # 101, Humble, TX 77396, USA. Tel: (713) 441-2111. Parts.

MAR-K Specialized Manufacturing, 6625 W Wilshire, Oklahoma City, OK 73132, USA. Tel: (405) 721-7945; fax: (405) 721-8906. Beds, customizing and trim parts.

Mid-America Designs Inc, PO Box 1368, Dept S95, Effingham, IL 62401, USA. Tel: (800) 500-8388; fax: (217) 347-2852. Accessories, parts.

Morrison Motor Co, 1170 Old Charlotte Road, Concord, NC 28025, USA. Tel: (704) 782-7716; fax: (704) 788-9514. Cars.

Mr Corvette Parts & Sales Inc, 2850 Bacon's Bridge Road, Summerville, SC 29485, USA. Tel: (803) 871-4043. Accessories, new, repro and used parts.

Old Air Products, 3056 SE Loop 820, Fort Worth, TX 76140, USA. Tel: (817) 551-000602; fax: (817) 568-0037. Air conditioning.

Pacifica Motoring Accessories, PO Box 2360, Atascadero, CA 93423, USA. Tel: (800) 488-7671; fax: (805) 466-4782. Accessories, parts.

M Parker Autoworks Inc, 374 N Cooper Road, Unit C-7, Dept HVAA, Berlin, NJ 08009, USA. Tel: (609) 753-0350; fax: (609) 753-00353. Battery cables, harnesses.

JT Piper's Auto Specialties Inc, # 2 Water Street, Box 140, Vermilion, IL 61955, USA. Tel: (800) 637-6111, orders (217) 275-3743; fax: (217) 275-3515. Parts.

John E Pirkle, 3706 Merion Drive, Augusta, GA 30907, USA. Tel: (706) 860-9047. Electrical parts.

Jack Podell Fuel Injection Specialist, 106 Wakewa Avenue, South Bend, IN 40617, USA. Tel: (219) 232-6430; fax: (219) 234-8632. Fuel system parts and rebuilding.

Dennis Portka, 4326 Beelow Drive, Hamburg, NY 14075, USA. Tel: (716) 649-0921. Horns, knock-off wheels.

Hugo Prado Limited Edition Corvette Art Prints, 3323 W Berteau Avenue, Chicago, IL 60618-2305, USA. Tel/fax: (800) 583-7627. Fine art prints.

Proteam Corvette Sales Inc, PO Box 606, Napoleon, OH 43545, USA. Tel: (419) 592-5086; fax: (419) 592-4242. Cars.

Repro Parts Manufacturing, PO Box 3690, San Jose, CA 95156, USA. Tel: (408) 923-2491. Parts.

Research Project 1956/1957, Michael Hunt, PO Box 5154, Madison, WI 53705, USA. Information.

Rik's Unlimited, 3758 Highway 18 S, Morganton, NC 28655, USA. Tel: (704) 433-6506; fax: (704) 437-7166. Accessories, parts.

Mary Jo Rohner's 1953-1962 Corvette Parts, 16367 Martincolt Road, Poway, CA 92064, USA. Tel: (619) 451-1933; fax: (619) 451-1999. Parts.

Rowley Corvette Supply Inc, 357 Main Street, Rowley, MA 01969, USA. Tel: (508) 948-7730; fax: (508) 948-3759. Parts, repairs, restoration.

Specialty Auto, 5409 S 70th Street, Omaha, NE 68117, USA. Tel: (402) 593-7339. Parts, repairs, service.

Stainless Steel Brake Corp, 11470 Main Road, Clarence, NY 14031, USA. Tel: (800) 448-7722, (716) 759-8666 in NY; fax: (716) 759-8688. Brake accessories and fluid, disc brakes, parking brake kits.

Still Cruisin' Corvettes, 5759 Benford Drive, Haymarket, VA 22069, USA. Tel: (703) 754-1960; fax: (703) 754-1222. Appraisals, repairs, restoration.

Verl Custom Tailor, Paul Becker, 424 Ward Parkway, Kansas City, MO 64112, USA. Tel: (816) 531-6200. Corvette car covers.

Virginia Vettes Parts & Sales, 110 Maid Marion Pl, Williamsburg, VA 23185, USA. Tel: (804) 229-0011. Interiors, parts.

Alan Whittier, 68 Scoble Pond Road, Derry, NH 03038, USA. Tel: (603) 437-3803. Parts.

Wild Bill's Corvette & Hi-Performance Center Inc, 451 Walpole Street, Norwood, MA 02062, USA. Tel: (617) 551-8858. Parts, rebuilding service.

R L Williams Classic Corvette Parts, PO Box 3307, Palmer, PA 18043, USA. Tel: (610) 258-2028; fax: (610) 253-6816. Literature, parts.

Vibratech Inc (Fluidampr), 537 E Delavan Avenue, Buffalo, NY 14211, USA. Tel: (716) 895-5404; fax: (716) 895-7258. Performance parts.

APPENDIX A
Corvette milestones 1953-1988

1953 Corvette introduced, Zora Arkus-Duntov joins Chevrolet.

1954 Work begins on OHV V8 engine.

1955 Duntov-prepared Corvette exceeds 150mph at Daytona Beach, Florida.

1956 First major design change with second-generation cars; optional hardtop available.

1957 Fuel injection produces one horsepower per cubic inch on 283cid small-block engine; Positraction rear axle available with 3.70, 4.11 or 4.56:1 ratios; heavy-duty suspension offered; five optional 283cid V8s offered with horsepower from 245 to 283.

1958 First full year of Auto Manufacturer's 'ban' on factory-sponsored competition; top Corvette engine now 290bhp.

1959 Metallic brake linings offered.

1960 First year for aluminium heads and radiator, top bhp now 315 via fuel injection.

1961 Aluminium radiator standard; direct-flow exhaust system offered as a no-cost option.

1962 New 327cid engine introduced with up to 360bhp.

1963 Third design generation arrives with Sting Ray split-window coupe and roadster; sintered metallic brakes optional; off-road exhaust system offered; RPO Z06 performance package offered for coupes only.

1964 One-piece rear window replaces split-window on coupe; top engine option 375bhp; transistorized ignition available.

1965 Four-wheel disc brakes standard; M22 four-speed close-ratio Heavy-Duty gearbox optional; telescopic steering column offered.

1966 Top bhp 425 via Turbo-Jet 427cid V8; fuel injection dropped.

1967 Wheel width increased to 6in, top bhp 435.

1968 Fourth design generation arrives, Sting Ray name temporarily dropped; wheel width increased to 7in; Turbo-Hydra-matic transmission optional.

1969 350cid V8 introduced, Stingray (one word) name returns; wheel width 8in; steering column lock standard, steering wheel diameter reduced to 15in.

1970 Turbo-Jet 454cid engine introduced, top bhp 390 on LS5, 460bhp available for competition on LS7; 350cid small-block V8 introduced.

1971 ZR1 factory racing option available with 330bhp 350cid engine; ZR2 option available with 425bhp 454cid engine.

1972 Engine output now SAE nett rather than SAE gross; top engine option is LS5 454cid with 270bhp; anti-theft alarm system standard.

1973 Energy-absorbing front bumper introduced; coupe's rear window fixed; LT1 engine dropped; L82 engine with 270bhp available; top engine is LS4, 454cid, 275bhp.

1974 Last year for dual exhausts and 454cid engines; rear end restyled to accommodate 5mph crash bumpers.

1975 Catalytic converters added. Last year for roadster; sole engine option is L82, rated at 205bhp; high-energy ignition system introduced.

1976 L82 engine rated at 210bhp; aluminium wheels offered.

1977 Leather seats standard; wipe/wash/headlamp dipper switches moved to steering column; power steering and brakes standard.

1978 Minor styling alteration produces fastback roof; limited-production Silver Anniversary and Indy Pace Car replica available; wiper control moved back to dash!; L82 engine option now 220bhp.

1979 60-series radial tyres offered; L82 rated at 225bhp; new lightweight bucket seats introduced (first used on Indy Pace Car in 1978).

1980 Aerovette *not* introduced; front and rear spoilers integrated in long-running fourth-generation body; 305cid California engine introduced for that state only; kerb weight reduced by 250lb.

1981 New glass-fibre-reinforced plastic monoleaf rear spring adopted on models with automatic transmission. Thinner side glass, stainless steel exhaust manifolds, lighter engine/interior materials to cut weight; quartz clock and six-way power seats standard.

1982 Drive-train for 1984 model introduced; L83 engine features dual throttle-body fuel injection (TBI), called 'Cross-Fire Injection' by Chevrolet. Four-speed overdrive automatic is only gearbox. Collector Edition model features an opening rear hatch.

1983 Fifth-generation Corvette makes debut, 10in shorter, 500-600lb lighter, with smoother, more aerodynamic bodywork, more glass, alloy wheels and ultra-low-profile tyres.

1986 Corvette revives the convertible and 7,264 are built for this model year. Major reinforcement of the convertible body results in the tightest new-generation car to date; similar improvements made to the coupe for 1987. Z51 Handling Suspension dropped.

1987 New Handling Suspension developed as RPO Z52 and is much improved from its forebear. Chevrolet builds over 10,000 Corvette convertibles for the first model year since 1969.

APPENDIX B
Technical specifications and major options

1953
Engine: Overhead-valve cast-iron six, 235.5cid, bore and stroke 3.56 × 3.93in, compression ratio 8:1, 150bhp at 4,200rpm.
Chassis: Box-section ladder-type. Ifs via coil springs and wishbones, live rear axle with leaf springs. Wheelbase 102in, overall length 167in, track 57/59in front/rear, tyres 6.70 × 15in.
Transmission: Standard Powerglide two-speed automatic with floor shift.
Options: Signal-seeking AM radio $145, heater-demister $91, whitewall tyres $25.

1954
Engine: As per 1953; new camshaft increased power to 155bhp in mid-model year.
Chassis: As per 1953.
Options: As per 1953, plus windscreen washer $12 and parking brake alarm $6.

1955
Engine: As per 1953, but few six-cylinder engines fitted. Most cars used optional ($135) overhead-valve cast-iron V8, 265cid, bore and stroke 3.75 × 3.00in, CR 8:1, 195bhp at 5,000rpm.
Chassis: As per 1953.
Options: As per 1954.

1956
Engine: Overhead-valve cast-iron V8, 265cid, bore and stroke 3.75 × 3.00in, CR 9.25:1, 210bhp at 5,200rpm (225bhp $175, 240bhp $160).
Chassis: Wheelbase 102in, overall length 168in, track 57/59in front/rear, tyres 6.70 × 15in.
Transmission: Three-speed manual gearbox (Powerglide $175).
Options: Power top $100, power windows $60, windscreen washer $11, detachable hardtop $200, signal-seeking AM radio $185, heater-demister $115, whitewall tyres $25.

1957
Engine: Overhead-valve cast-iron V8, 283cid, bore and stroke 3.87 × 3.00in, CR 9.5:1, 220bhp at 4,800rpm (245bhp $140, 270bhp $170, 250bhp $450, fuel-injection 283bhp $450 or $675 with cold-air induction system).
Chassis: As per 1956, with optional Positraction $45 and racing suspension $725.
Transmission: Three-speed manual gearbox (automatic $175, four-speed gearbox $188).
Options: Signal-seeking AM radio $185, detachable hardtop $215, power top $130, courtesy light package $8, 15 × 5.5in wheels $14, heater-demister $118, windscreen washer $12, parking brake alarm $5, whitewall tyres $32, dual carburation $151, two-tone paint $19, Motorola radio $125, electric windows $55.

1958
Engine: As per 1957 except 230bhp at 4,800rpm (245bhp $150, f-i 250bhp $484, f-i 290bhp $484).
Chassis: Wheelbase 102in, overall length 177.2in, track 57/59in front/rear, tyres 6.70 × 15in. Positraction $48, heavy-duty brakes and suspension $780, available final drive ratios 3.70, 4.11, 4.56:1.
Transmission: Three-speed manual gearbox (Powerglide $188, four-speed manual $215).
Options: Power top $140, heater-demister $97, extra cove colour $16, detachable hardtop $215, signal-seeking AM radio $144, power windows $59, 15 × 5.5in wheels no extra charge, windscreen washer $16, whitewall tyres $32, courtesy lights $6, parking brake alarm $5.

1959
Engine: As per 1958 (245bhp $151, 270bhp $183, 250bhp $484, 290bhp $484).
Chassis: As per 1958 (Positraction $48, HD brakes and suspension $425).
Transmission: As per 1958.
Options: Power top $140, windscreen washer $16, transistor radio $150, deluxe heater $102, two-tone paint $16, electric windows $59, courtesy lights $6, parking brake alarm $5, sunshades $11, 15 × 5in wheels no extra charge, detachable hardtop $237.

1960
Engine: As per 1959 except CR 9.25:1 (245bhp $151, 270bhp $183. f-i 275/315bhp $484).
Chassis: As per 1959. Positraction $43, HD brakes and suspension $333. Available final-drive ratios 3.70, 4.11, 4.56:1.
Transmission: Three-speed manual gearbox (automatic $199, four-speed $188).
Options: Power top $140, windscreen washer $16, signal-seeking transistor radio $138, deluxe heater $102, detachable hardtop $237, two-tone paint $16, electric windows $60, whitewall tyres $32, courtesy lights $6, parking brake alarm $5, sunshades $11, special 15 × 5.5in wheels no extra charge.

1961
Engine: As per 1960 except CR 9.5:1 (245bhp $151, 270bhp $183, f-i 275/

315bhp $484).
Chassis: As per 1959 (options as per 1959 plus metallic brakes $38).
Transmission: As per 1960.
Options: Power top $161, windscreen washer $16, signal-seeking transistor radio $138, deluxe heater $102, detachable hardtop $237, two-tone paint $16, electric windows $59, whitewall tyres $32, blackwall nylon tyres $5, positive crankcase ventilation $5, 24-gallon (US) fuel tank $161, special 15 × 5.5in wheels no extra charge.

1962
Engine: Overhead-valve cast-iron V8 327cid, bore and stroke 4.00 × 3.25in, CR 10.5:1, 250bhp at 4,400rpm (300bhp $54, 340bhp $108, f-i 360bhp $484).
Chassis: As per 1961, direct flow exhaust system no extra charge, metallic brake linings $38, Positraction $43, HD brakes and suspension $333.
Transmission: Three-speed manual gearbox (automatic $199, four-speed $188).

1963
Engine: As per 1962 (f-i 360bhp $430).
Chassis: Steel perimeter frame with independent suspension front and rear, the latter via three-link with double-jointed open drive-shafts at either side, control arms and trailing radius rods with single transverse leaf spring. Wheelbase 98in, overall length 175.2in, track 56.8/57.6in front/rear, tyres 6.70 × 15in. Sintered metallic brake linings $38, off-road exhaust system $38, RPO Z06 performance package for coupe (including metallic power brakes, HD shock absorbers, stabilizer bars, knock-off aluminium wheels, Positraction, four-speed, 360bhp engine) $1,818, Positraction $43. Available final-drive ratios 3.08, 3.36, 3.55, 3.70, 4.11, 4.56:1.
Transmission: Three-speed manual gearbox (automatic $199, four-speed $188).
Options: Power brakes $43, power steering $73, air conditioning $422, hardtop $237, signal-seeking transistor radio $138, electric windows $60, whitewall tyres $32, blackwall Nylon tyres $16, HD brakes with metallic linings $38, Sebring Silver paint $81, Woodgrain steering wheel $15, aluminium knock-off wheels $323, AM-FM radio $174, tinted windscreen $11, tinted glass $16, leather seat trim $81.

1964
Engine: As per 1963 (300bhp $54, 365bhp $108, f-i 365bhp $538).
Chassis: As per 1963.
Transmission: As per 1963.
Options: As per 1963 plus 36-gallon (US) fuel tank for coupe $202, reversing lamps $11.

1965
Engine: As per 1963 (other 327 engines: 300bhp $54, 350bhp $108, 365bhp $129, f-i 375bhp $538; L78 396cid 425bhp $292).
Chassis: As per 1963 except standard tyres now 7.75 × 15in, four-wheel disc brakes standard, special front and rear suspension $38, Positraction $43, side-mounted exhaust system $135.
Transmission: As per 1963 (special HD close-ratio four-speed gearbox $237).
Options: As per 1963 plus teakwood steering wheel $48, AM-FM radio with power antenna $237, 'goldwall' tyres $51, telescopic steering column $43.

1966
Engine: As per 1963 except 300bhp at 5,000rpm (350bhp $105, L39 427cid, 390bhp $181, L72 427cid, 425bhp $312, transistor ignition system $73). Fuel-injected engines dropped.
Chassis: As per 1965.
Transmission: As per 1965.
Options: As per 1965 plus traffic hazard lamp switch $12.

1967
Engine: As per 1966 (350bhp $105, L36 427cid 390bhp $200, L68 427cid 400bhp $306, L71 427cid 435bhp $437, aluminium cylinder heads for L71 $368.
Chassis: As per 1965. Available final-drive ratios 3.08, 3.36, 3.55, 3.70, 4.11:1.
Transmission: As per 1966.
Options: Power brakes $42, power steering $95, air conditioning $413, front shoulder belts $26, 36-gallon (US) fuel tank for coupe $198, tinted windows $16, tinted windscreen $11, headrests $42, heater-demister $98, power windows $58, AM-FM radio $173, black vinyl roof cover $53, leather upholstery $79, speed warning indicator $11, telescopic steering wheel $42, four-ply whitewall tyres $31, red stripe Nylon tyres $47, hardtop $231.

1968
Engine: As per 1967.
Chassis: As per 1967, wheelbase 98in, overall length 182.5in, track 58.7/59.4in front/rear, tyre size F70—15. Available rear axle ratios 2.73, 3.08, 3.36, 3.55, 4.11:1.
Transmission: As per 1967 (HD gearbox $263, Turbo-Hydra-matic $226).
Options: Power brakes $42, power steering $95, air conditioning $413, deluxe shoulder belts $25, rear window demister $32, tinted windows $16, tinted windscreen $11, head restraints $42, HD power brakes $384, power windows $58, AM-FM radio $173, with stereo $278, black vinyl roof $53, leather seats $79, speed warning indicator $11, adjustable steering wheel $42, detachable hardtop $232, wheel covers $58, red stripe F70—15 tyres $31, white stripe tyres $32, alarm system $26.

1969
Engine: Overhead-valve V8, 350cid, bore and stroke 4.00 × 3.48in, CR 10.25:1, 300bhp at 4,800rpm (L36 427cid 400bhp $326, L71 427cid 435bhp $437, L89 427cid, 435bhp with aluminium heads $832).
Chassis: As per 1968, steering column lock standard.
Transmission: As per 1968 (HD gearbox $290).
Options: As per 1968.

1970
Engine: As per 1969 (350bhp $158, LT1 350cid 370bhp $448, LS5 454cid 390bhp $290, LS7 454cid 460bhp $3,000).
Chassis: As per 1969. Transistor ignition $64, special suspension $29, Positraction $12, HD clutch $63. Available final-drive ratios 2.73, 3.08, 3.36, 3.55, 4.11, 4.56:1.
Transmission: Four-speed manual gearbox (automatic no extra charge, close-ratio four-speed no extra charge, HD four-speed $95).
Options: As per 1969.

1971
Engine: Overhead-valve cast-iron V8, 350cid, bore and stroke 4.00 × 3.48in, CR 8.5:1, 270bhp at 4,800rpm (350bhp $483, LS5 454cid 365bhp $295, LS6 454cid 425bhp $1,221).
Chassis: As per 1969 (ZR1 package: HD brakes, close-ratio four-speed, special front stabilizer bar, special springs and shock absorbers, transistorized ignition and LT1 engine $1,010; ZR2: all ZR1 except LS6 engine $1,747), side-mounted exhaust system $117.
Transmission: As per 1970 (HD gearbox $100).
Options: Power brakes $48, power steering $116, air conditioning $465, alarm system $32, HD battery $16, deluxe shoulder belts $42, rear window demister $42, AM-FM radio $178, with stereo $283, black vinyl roof cover $63, tilt steering wheel $84, white stripe tyres $30, white letter tyres $44, custom trim $158, custom wheel cover $63, power windows $85.

1972
Engine: As per 1971 except 200bhp at 5,500rpm (255bhp $483, LS5 454cid 270bhp $295).
Chassis: As per 1969. ZR1 package with 255bhp $1,010.
Transmission: As per 1971. Automatic no extra charge with standard engine, $97 extra with others, not available with 454; close-ratio four-speed no extra charge.
Options: As per 1971 except alarm system standard.

1973
Engine: As per 1972 except 190bhp (net) at 4,400rpm (250bhp $299, LS4 454cid 275bhp $250).
Chassis: As per 1972 plus energy-absorbing front bumper. Off-road suspension and brake package $369.
Transmission: As per 1972.
Options: As per 1972 plus cast-aluminium bolt-on wheels $175.

1974
Engine: Overhead-valve cast-iron V8 350cid, bore and stroke 4.00 × 3.48in, CR 9:1, 250bhp at 5,200rpm (250bhp $299, LS4 454cid 270bhp $250).
Chassis: Wheelbase 98in, overall length 185.5in, tyres GR70–15 (off-road suspension and brake package $400, Gymkhana suspension $7).
Transmission: As per 1973.
Options: power brakes $49, power steering $117, air conditioning $467, custom interior $154, power windows $83, custom shoulder belts $41, detachable hardtop $267, vinyl-covered hardtop $329, rear window demister $41, tilt steering column $82, white stripe radial tyres $32, white letter radial tyres $45, dual horns $4, AM-FM stereo radio $276, AM-FM radio $173, HD battery $15, map light $5, cast-aluminium wheel trim $175.

1975
Engine: As per 1974 except CR 8.5:1, 165bhp at 3,800rpm, catalytic convertor, High Energy ignition system (205bhp $335).
Chassis: As per 1974; suspension options deleted.
Transmission as per 1974.
Options: As per 1974; cast-aluminium wheel trim deleted.

1976
Engine: As per 1975 except 185bhp (220bhp $335).
Chassis: As per 1975.
Transmission: As per 1975.
Options: As per 1975 except aluminium wheels now offered at $321.

1977
Engine: As per 1976 (210bhp $495).
Chassis: As per 1976. Tyres GR70–15B.
Transmission: As per 1976.
Options: AM-FM stereo radio $281, with tape deck $414, aluminium road wheels $321, power windows $100, glass canopy roof $200, speed control, tilt steering wheel $165, air conditioning $450. (Leather seats, power steering and power brakes now standard.)

1978
Engine: As per 1977 (220bhp $525).
Chassis: As per 1977.
Transmission: As per 1977.
Options: AM-FM stereo radio $220, with tape deck $439, with CB $525, aluminium road wheels $345, tilt steering wheel $178, removable glass roof panels $349, air conditioning $450; also power windows, rear window demister, speed control, power door locks.

1979
Engine: As per 1978 except 195bhp (225bhp $595).
Chassis: As per 1978.
Transmission: As per 1978.
Options: As per 1978.

1980
Engine: As per 1979 (230bhp $595), also overhead-valve cast-iron V8, 305cid, bore and stroke 3.73 × 3.48in, 180bhp for California.
Chassis: As per 1979.

Transmission: As per 1979.
Options: As per 1979.

1981
Engine: As per 1980.
Chassis: As per 1980, transverse glass-fibre monoleaf rear spring on models with automatic.
Transmission: As per 1980.
Options: As per 1978 plus roof-mounted luggage carrier $144.

1982
Engine: Overhead-valve cast-iron V8 with Throttle-Body ('Cross-fire') fuel injection, 350cid, bore and stroke 4.00 × 3.48in, 200bhp at 4,300rpm.
Chassis: As per 1981.
Transmission: Three-speed automatic only.
Options: Aluminium road wheels $458, power seats $197, power door locks $155, luggage rack $144, cruise control $165, removable roof panels $443, custom paint $438; also rear window demister, AM-FM stereo with tape and/or CB.

1984
Engine: Overhead-valve cast-iron V8, 350cid, bore and stroke 4.00 × 3.48in, 205bhp at 4,300rpm.
Chassis: Unequal-length A-arms, transverse glass-fibre leaf spring, tubular shock absorbers and anti-sway bar front; upper and lower trailing arms, lateral arms, tie rods, half-shafts, transverse glass-fibre leaf spring, tubular shock absorbers and anti-sway bar rear; 11.5in vented disc brakes front and rear, vacuum-assisted; cast-alloy road wheels, 15in standard, 16in optional; rack-and-pinion power-assisted steering, overall ratio 13:1, two turns lock-to-lock. Wheelbase 96in, overall length 176.5in, overall width 71in, height 46.9in.
Transmission: Four-speed manual gearbox with computer-controlled overdrive standard.
Options: To include usual Corvette equipment plus electrically adjustable rear-view mirrors, Z51 handling package (HD suspension, aluminium prop-shaft and half-shafts, 16in wheels and tyres. Estimated base price FOB Detroit $24,000.

1985
Engine: 230bhp at 4,000rpm; new heavy-duty cooling system.
GM multi-port fuel injection.
Chassis: Cast-alloy road wheels, 9.5in front and rear; gas-pressurized Delco-Bilstein shock absorbers. Heavy-duty 8.5in ring-gear differential in manual-shift models.

1986
Chassis: New convertible model on individual chassis with 9.5in wheels, lower recommended tyre pressures, 10mm higher ride height, deflected-disc Delco shock absorbers.

1987
Engine: 240bhp at 4,000rpm, compression ratio 9.5:1, torque 345lb/ft at 3,200rpm, roller-type lifters, raised-rail rocker-arm covers, relocated spark plugs.

APPENDIX C
Serial numbers, production, weight and base prices

Year	Model	Prefix	Starts	Ends	Produced	Wgt (lb)	Price ($)
1953	all	E53F	-001001	-001300	300	2,705	3,513
1954	all	E54S	-001001	-004640	3,640	2,705	3,523
1955	six	E55S	-001001	-001700	6	2,650	2,799
	V8	VE55S			668	2,620	2,934
1956	all	E56S	-001001	-004467	3,388	2,764	3,149
1957	all	E57S	-100001	-106339	6,246	2,730	3,465
1958	all	J58S	-100001	-109168	9,168	2,793	3,631
1959	all	J59S	-100001	-109670	9,670	2,840	3,875
1960	all	00867S	-100001	-110261	10,261	2,840	3,872
1961	all	10867S	-100001	-110939	10,939	2,905	3,934
1962	all	20867S	-100001	-114531	14,531	2,925	4,038
1963	coupe	30837S	-100001	-121513	10,594	2,859	4,252
	conv	30867S			10,919	2,881	4,037
1964	coupe	40837S	-100001	-122229	8,304	2,960	4,252
	conv	40867S			13,925	2,945	4,037
1965	coupe	194375S	-100001	-123562	8,186	2,975	4,321
	conv	194675S			15,376	2,985	4,106
1966	coupe	194376S	-100001	-127720	9,958	2,985	4,295
	conv	194676S			17,762	3,005	4,084
1967	coupe	194377S	-100001	-122940	8,504	3,000	4,353
	conv	194677S			14,436	3,020	4,141
1968	coupe	194378S	-400001	-428566	9,936	3,055	4,663
	conv	194678S			18,630	3,065	4,320
1969	coupe	194379S	-700001	-738762	22,154	3,140	4,781
	conv	194679S			16,608	3,145	4,438
1970	coupe	194370S	-400001	-417316	10,668	3,184	5,192
	conv	194670S			6,648	3,196	4,849
1971	coupe	194371S	-100001	-121801	14,680	3,153	5,536
	conv	194671S			7,121	3,167	5,299
1972*	coupe	1Z37K2S	-500001	-520486	20,486	3,215	5,472
	conv	1Z67K2S	-500001	-506508	6,508	3,215	5,246
1973	coupe	1Z37K3S	-400001	-424372**	24,372**	3,407	5,921
	conv	1Z67K3S	-400001	-406093**	6,093**	3,407	5,685
1974	coupe	1Z37J4S	-400001	-432028	32,028	3,532	6,372
	conv	1Z67J4S	-400001	-405474	5,474	3,532	6,156
1975	coupe	1Z37J5S	-400001	-433836	33,836	3,532	7,117
	conv	1Z67J5S	-400001	-404629	4,629	3,532	6,857
1976	coupe	1Z37J6S	-400001	-446558	46,558	3,445	7,605
1977	coupe	1Z67J7S	-100001	-149213	49,213	3,448	8,648
1978	coupe	1Z67J8S	-100001	-147667	47,667	3,401	9,645
1979	coupe	1Z67J9S	-400001	-453807	53,807	3,372	12,313
1980	coupe	*	-100001	-140614	40,614	3,206	13,965
1981	coupe	*	-100001	-145631	45,631	3,179	16,259
1982	coupe	*	-100001	-125407	25,407	3,232	18,920
1984	coupe	*	-100001	-151547	51,547	3,200	21,800
1985	coupe	*	-100001	-139729	39,729	3,191	24,878
1986	coupe	*	-100001	-127673	27,673	3,216	27,502
	conv	*	-100001	-107264	7,264	3,325	32,507
1987	coupe	*	-100001	-120007	20,007	3,200	28,474
	conv	*	-100001	-110625	10,625	3,325	33,647
1988	coupe	*	-100001	-up	n.a.	n.a.	n.a.
	conv	*	-100001	-up	n.a.	n.a.	n.a.

*The new numbering system which took effect in 1972 designates the following, using prefix 1Z37K2S as an example: 1 = Chevrolet Division, Z = Corvette, 37 = coupe (67 = convertible), K = base engine ('J' from 1974, 'Y' for 454s), 2 = model year (1972), and S = St Louis assembly plant. The system which took effect in 1980 follows a similar format:
1980: 1 (Model) ( ) A ( ) 100001 up
1981: ( ) G1 ( ) Y87L B ( ) 100001 up
1982: ( ) G1 ( ) (Model) 8 C ( ) 100001 up
Serial number spans are not helpful in determining production in later years, because factory records indicate only the starting and ending numbers selected; however, the highest number has been applied to the 'Ends' column in accord with the production figure.

**Proportional estimates from total model year production of 30,465.

APPENDIX D

How fast? How economical? How heavy? Performance figures for Corvettes

In this section of MRP *Collector's Guides,* my friend and colleague Graham Robson traditionally notes: 'I do not believe in quoting from factory handouts. Nor, for that matter, do I trust the enthusiastic figures quoted by some of the fringe magazines.' I couldn't agree more. This selection of Corvette performance figures has accordingly been culled from responsible journals, and wherever I had a choice I usually chose *Road & Track. R&T's* editors are often accused of driving with marshmallow feet compared to other less staid and noisier publications — but maybe this comes out of respect for the borrowed machinery. I will never forget the Triumph Stag I saw after a road test by one of *R&T's* rivals. Somehow they had managed to change the wheelbase by two full inches.

The selection has been chosen to give a broad cross-reference to Corvette performance over the years. There are very few symbols or codes. An asterisk (*) indicates a fuel-injected engine. A double asterisk (**) indicates automatic transmission. The letter 'e' following any figure means the figure was estimated — either by the original testers or by myself (when the exact figure wasn't quoted, as in acceleration graphs).

The traditional mid-range acceleration figures of British testers, using top gear, are notably absent. I wanted to stick to as few magazines as possible, and the plethora of Corvettes tested by the American publications made their use essential; but few of them quote 50-70mph-in-top or like statistics.

At the foot of each column you'll find the month, year and publication of the road test, which will make it easy to look up the original should you need more information. The magazine symbols are MT *(Motor Trend),* R&T *(Road & Track,* SCI *(Sports Cars Illustrated),* C/D *(Car and Driver,* which succeeded SCI) and SCG *(Sports Car Graphic),* CG *(Consumer Guide New Car Annual).*

The performance headings begin with displacement in cubic inches (CID) rather than litres, for the same reason I used cubic inches in the previous pages — familiarity to the bulk of Corvette enthusiasts. If you must know how many litres in any CID, simply divide by 61. Brake horsepower (BHP) is SAE gross through to the 1970 model, and afterwards SAE nett. The gross bhp figures largely exaggerate what you could actually record at the rear wheels with a dyno, while the nett may be a shade pessimistic, but in general much more accurate. The column labelled 'Axle' refers of course to the final-drive ratio, which is crucial in the case of Corvettes; as you will see, it made quite a difference. One other caveat for British readers: the mpg figures are in American gallons; to obtain the Imperial gallon figure, divide by four and multiply by five. (By and large, it will still be pretty low!)

It's hardly necessary to tell you what you're going to observe readily yourself, but I'm struck by a few interesting statistics among the following. The first was the difference a V8 engine made (1956 compared with 1954). Yet the Corvette Six wasn't a slouch, considering that the 1956 V8 had 50 per cent more horsepower. The second is how little difference those incremental power gains seem to be. Compare the 270bhp carburettor-equipped 283 of 1957 with the 290bhp 'fuelie'. The 270 was actually quicker in the 0-60mph sprint and quarter-mile time, while turning in the same 95mph at the end of the quarter-mile. Of course, we are dealing with different cars and road testers.

The third point is one that confounds much of my previous thinking about the age of Federal regulation of motor cars. It is remarkable how little performance actually dwindled betweem 1968 and 1979. In 1968 a 427 Corvette weighing 3,425lb did 0-60mph in 6.3 seconds and the standing quarter-mile in 14.1 seconds; in 1979 a 350 Corvette weighing 230lb *more* did 0-60mph in 6.6 seconds and the standing quarter-mile in 15.6 seconds; and they both had a 3.55:1 final-drive ratio. (The difference, of course, was at the top end, with the 427 pegged at 160mph, but even Corvette drivers don't ordinarily use that portion of the power curve too often.)

Things became relatively bleaker in 1982, when the 350 was rated at only 200bhp nett. But even then it was capable of 125mph, with 0-60mph in under 8 seconds — at a gain of almost 10mpg in fuel economy. So maybe Corvette performance didn't suffer as badly in the 1970s as we've tended to assume.

	1954	1956	1957	1957	1958	1958	1958	1959	1959	1960	1961
CID	235.5	265	283	283	283	283	283	283	283	283	283
BHP	150	225	283*	270	230	250	290*	250	290*	270	315*
Axle	3.55	3.55	3.70	3.70	4.11	3.70	4.11	3.70	4.11	n.a.	3.70
Weight (lb)	2,890	2,980	2,840	2,960	2,900e	2,912	2,900e	2,950e	3,020	2,950e	3,065
Mean maximum (mph)	107.1	120.0	125.0	122.5	103.1	125.0	118.7	120.0	128.0	n.a.	140.0e
Accleration (sec)											
0-30mph	3.7	3.4	2.4	3.1	—	3.3	—	—	3.1	—	2.6
0-40mph	5.3	4.6	3.4	3.8	—	4.5	—	—	4.2	—	3.8
0-50mph	7.7	6.0	4.9	4.9	—	5.8	—	—	5.1	—	4.6
0-60mph	11.0	7.5	6.6	6.8	9.2	7.6	6.9	7.8	6.6	8.4	6.0
0-70mph	14.8	10.0	8.8	8.7	—	9.5	—	—	7.5	—	7.9
0-100mph	—	19.3	18.2	17.6	—	21.4	—	—	15.5	—	14.2
¼-mile (sec)	18.0	15.9	14.2	15.0	17.4	15.7	15.6	15.7	14.5	16.1	15.5
¼-mile (mph)	76.0	91.0	93.0	95.0	83.3	90.0	95.0	90.0	96.0	89.0	106.0
Direct top gear (sec)											
50-70mph	—	4.5	—	—	50-80:	—	50-80:	50-80:	—	—	—
60-80mph	—	4.7	—	—	9.8	—	5.9	6.8	—	—	—
Overall fuel consumption (mpg — US)	18.0	12.0	12.0	10.5	12.9	18.5	13.9	14.6	11.7	17.0e	14.0
Original test	6.54	5.56	6.57	6.57	3.58	12.57	3.58	4.59	1.59	7.60	12.60
Publication	R&T	SCI	SCI	SCI	MT	SCI	MT	MT	R&T	MT	SCI

	1961	1962	1963	1963	1964	1965	1965	1966	1967	1968	1968
CID	283	327	327	327	327	327	396	427	327	327	327
BHP	230	360*	360*	300	300	375*	425	425	300	350	350
Axle	3.70	3.70	3.70	3.36	3.36**	3.70	3.70	4.11	3.36	3.70	3.70
Weight (lb)	3,000e	3,035	3,030	3,180	3,050	3,050	3,260	3,360	3,160	3,260	3,445
Mean maximum (mph)	n.a.	150.0e	142.0	118.0	130.0	138.0	136.0	135.0	121.0	128.0	117.0
Accleration (sec)											
0-30mph	3.8	2.9	2.5	2.2	3.2	2.9	3.1	2.5	3.4	4.0	3.2
0-40mph	—	3.8	3.4	3.1	4.4	3.9	3.8	—	4.6	5.0	4.8e
0-50mph	—	4.8	4.5	4.2	6.1	5.2	4.8	4.2	5.9	6.1	—
0-60mph	8.3	5.9	5.9	6.1	8.0	6.3	5.7	5.6	7.8	7.7	7.1
0-70mph	—	7.5	8.0	7.0	10.5	7.8	7.5	—	10.0	9.7	—
0-100mph	—	13.5	16.5	14.5	20.2	14.7	13.4	—	23.1	20.7	—
¼-mile (sec)	—	14.5	14.9	14.5	15.2	14.4	14.1	13.4	16.0	15.6	15.0
¼-mile (mph)	—	104.0	95.0	100.0	85.0	99.0	103.0	105.0	86.5	92.0	92.0
Direct top gear (sec)											
50-70mph	—	—	—	—	6.0	2.4	2.5	2.0	—	3.6	—
60-80mph	—	—	—	—	—	—	—	—	—	—	—
Overall fuel consumption (mpg — US)	13.0e	n.a.	12.5	14.5	14.8	13.0	10.0	n.a.	15.4	13.0	15.7
Original test	9.61	12.61	11.62	4.63	3.64	12.64	8.65	3.66	2.67	1.68	3.68
Publication	MT	SCG	SCG	C/D	R&T	R&T	R&T	MT	R&T	R&T	MT

	1968	1969	1969	1970	1973	1974	1976	1977	1978	1979	1982	1984	1987
CID	427	350	427	454	350	350	350	350	350	350	350	350	350
BHP	435	300	435	390	250	250	210	210	220	220	200*	205	240
Axle	3.55	3.36**	4.11	3.01	3.70	3.70	3.55	3.70	3.70	3.55**	2.87**	3.31	3.07
Weight (lb)	3,425	3,505	3,262	3,740	3,520	3,490	3,610	3.540	3,490	3,655	3,425	3,200	3,325
Mean maximum (mph)	160.0	132.0	122.0	144.0	124.0	124.0	132.0	132.0	132.0	130.0e	125.0	140.0	150.0
Accleration (sec)													
0-30mph	2.7	—	2.5	2.8	3.1	3.3	3.4	2.6	2.3	2.5	2.7	2.9	2.8
0-40mph	3.8e	—	3.4	3.9	4.3	—	—	—	—	—	—	3.9	3.8
0-50mph	—	—	4.5	5.0	5.6	—	—	5.3	4.8	—	5.8	5.0	4.8
0-60mph	6.3	8.4	6.1	7.0	7.2	7.4	8.1	6.8	6.5	6.6	7.9	7.0	6.5
0-70mph	8.0e	—	—	8.9	9.1	—	—	—	—	—	10.6	—	—
0-100mph	—	21.1	14.8	14.0	17.9	18.5	23.6	18.0e	17.9	18.5e	24.8	—	—
¼-mile (sec)	14.1	16.0	14.3	15.0	15.5	15.8	16.5	15.5	15.2	15.6	16.1	15.5	15.0
¼-mile (mph)	103.0	82.0	98.0	93.0	94.0	92.5	87.0	92.5	95.0	91.0	84.5	88.0	93
Direct top gear (sec)													
50-70mph	2.9	—	—	—	—	—	—	—	—	—	—	—	—
60-80mph	—	—	—	—	—	—	—	—	—	—	—	—	—
Overall fuel cons (mpg — US)	13.6	14.3	10.0	9.0	14.5	13.5	14.0	15.0	15.0	12.0	21.5	18.0	17.0
Original test	3.68	6.69	3.69	9.70	6.73	2.74	3.76	6.77	4.78	4.79	11.82	1984	1986
Publication	MT	R&T	R&T	R&T	R&T	R&T	R&T	R&T	R&T	R&T	R&T	CG	CG

APPENDIX E
Interchangeability of Chevrolet small-block engines

Since 1955 the Chevrolet small-block engine has been produced in a variety of displacements from 262cid to 400cid. Variations in bore and stroke, crankshaft main journal and rod journal diameters and two- or four-bolt main bearing cap design are as follows:

Engine and model year	Bore (in)	Stroke (in)	Main journal dia(in)	Rod journal dia(in)	Main bearing cap
262 1975-76	3.671	3.1	2.45	2.1	2-bolt
265 1955-56	3.75	3.0	2.30	2.0	2-bolt
267 1979-81	3.50	3.48	2.45	2.1	2-bolt
283 1957-67	3.875	3.0	2.30	2.0	2-bolt
302 1967	4.0	3.0	2.30	2.0	2-bolt
302 1968-69	4.0	3.0	2.45	2.1	4-bolt
305 1976-81	3.736	3.48	2.45	2.1	2-bolt
307 1968-73	3.875	3.25	2.45	2.1	2-bolt
327 1962-67	4.0	3.25	2.30	2.0	2-bolt
327 1968-69	4.0	3.25	2.45	2.1	2-bolt
350 1967-81	4.0	3.48	2.45	2.1	2 & 4-bolt
400 1970-72	4.125	3.75	2.65	2.1	4-bolt
400 1973-80	4.125	3.75	2.65	2.1	2-bolt

In 1968 the 4in bore block was updated to incorporate four-bolt main bearing caps and increased thickness main bearing webs. The four-bolt caps were retained on all 302 and 350 engines until 1971, at which time the low-compression passenger car two-barrel and four-barrel carburettor engines were released again with two-bolt main caps. The 350cid engine for the Z-28 Camaro through to 1974, the LT-1 and L-82 Corvette through to 1978 and the LS-9 truck through to 1978 were manufactured with four-bolt main caps.

Cylinder blocks for the 262, 265, 267, 283, 305 and 307cid engines were never designed to accommodate four-bolt main caps and do not have the extra cast iron in the main bearing webs that is provided with the 4in bore blocks.